Study Guide and Mapping Workbook

C0-DAU-778

Vincent J. Del Casino, Jr.

California State University, Long Beach

THIRD EDITION

World Regions in Global Context
Peoples, Places, and Environments

Sallie A. Marston

Paul L. Knox

Diana M. Liverman

PEARSON

Prentice
Hall

Upper Saddle River, NJ 07458

Editor-in-Chief, Science: Nicole Folchetti
Publisher: Dan Kaveney
Associate Editor: Amanda Brown
Senior Managing Editor, Science: Kathleen Schiaparelli
Assistant Managing Editor, Science: Gina M. Cheselka
Project Manager: Jessica Barna
Supplement Cover Manager: Paul Gourhan
Supplement Cover Designer: Victoria Colotta
Senior Operations Supervisor: Alan Fischer
Director of Operations: Barbara Kittle
Photo Credit: Ladi Kirn/Alamy Images

© 2008 Pearson Education, Inc.

Pearson Prentice Hall

Pearson Education, Inc.

Upper Saddle River, NJ 07458

Printed in the United States of America

10   9   8   7   6   5   4   3   2

ISBN 13:   978-0-13-229805-6

ISBN 10:   0-13-229805-8

Pearson Education Ltd., *London*
Pearson Education Australia Pty. Ltd., *Sydney*
Pearson Education Singapore, Pte. Ltd.
Pearson Education North Asia Ltd., *Hong Kong*
Pearson Education Canada, Inc., *Toronto*
Pearson Educación de Mexico, S.A. de C.V.
Pearson Education—Japan, *Tokyo*
Pearson Education Malaysia, Pte. Ltd.

Table of Contents

Preface: Introduction to Mapping

Introduction

This workbook provides you with basic map reading skills and offers you exercises to further analyze the spatial dimensions of global and regional environmental, political-economic, and socio-cultural distributions, diffusions, and change. In this first chapter, you can take a short, self-administered exam to test your knowledge of mapping and map reading. After you have completed the short exam in the Preface, you are ready to take on the Mapping Exercises in the main chapters of the Study Guide. As a suggestion, you might consider removing each of the maps and photocopying them so that you have plenty of copies for studying later.

In addition, this preface, like all the chapters that follow, is designed to be used in conjunction with the Marston, Knox, and Liverman (2008) text *World Regions in Global Context: Peoples, Places, and Environments*. Throughout the guide, you will be referred to the main textbook for the data to be used in making your maps and for discussions related to the types of maps geographers use. The following introduction is broken down into three sections—understanding location, testing your map knowledge, and completing the exercises. The first section contains a number of key map concepts that will be valuable as you continue working through this workbook. This first section complements the Appendix in the text, which is titled "Maps and Geographic Information Systems."

Understanding Location[1]

Often, location is *nominal,* or expressed solely in terms of the names given to regions and places. We speak, for example, of Washington, D.C., or of Georgetown, a location within Washington, D.C. Location can also be used as an *absolute* concept, whereby locations are fixed mathematically through coordinates of latitude and longitude (**Map 1.1**). **Latitude** refers to the angular distance of a point on Earth's surface, measured in degrees, minutes, and seconds north or south from the equator, which is assigned a value of 0°. Lines of latitude around the globe run parallel to the equator, which is why they are sometimes referred to as *parallels*. **Longitude** refers to the angular distance of a point on Earth's surface, measured in degrees, minutes, and seconds east or west from the *prime meridian* (the line that passes through both poles and through Greenwich, England, and which is assigned a value of 0°). Lines of longitude, called *meridians,* always run from the North Pole (latitude 90° north) to the South Pole (latitude 90° south). Georgetown's coordinates are 38°55' N, 77°00' W (**see Marston et al. 2008, p. A-3, Figure A.3**).

Thanks to the Global Positioning System (GPS), it is very easy to determine the latitude and longitude of any given point. The **Global Positioning System** consists of 21 satellites (plus 3 spares) that orbit Earth on precisely predictable paths, broadcasting highly accurate time and locational information. The U.S. government owns the GPS, but the information transmitted by the satellites is freely available to everyone around the world. All that is needed is a GPS receiver. Basic receivers cost less than $100 and can relay latitude, longitude, and height to within 10 meters day or night, in all weather conditions, in any part of the world. The most precise GPS receivers, costing thousands of dollars, are accurate to within a centimeter. GPS has drastically increased the accuracy and efficiency of collecting spatial data. In combination with **geographic information systems** (GIS) and **remote sensing** (the collection of information about parts of Earth's surface by means of aerial photography or satellite imagery), GPS has revolutionized mapmaking and spatial analysis (see, for example, **Marston, Knox, and Liverman, 2007, p. 210, Fig. 5.2,** along with the satellite imagery presented at the beginning of all the chapters).

Location can also be *relative,* fixed in terms of site or situation. **Site** refers to the physical attributes of a location: its terrain, its soil, vegetation, and water sources, for example. **Situation** refers to the location of a place relative to other places and human activities: its accessibility to routeways, for example, or its nearness to population centers. Washington, D.C., has a low-lying, riverbank site and is situated at the head of navigation of the Potomac River, on the eastern seaboard of the United States.

Finally, location also has a *cognitive* dimension, in that people have cognitive images of places and regions, compiled from their own knowledge, experience, and impressions. **Cognitive images** (sometimes

---

[1]This section and the following section, "Understanding Maps," were originally part of Marston, Knox, and Liverman (2002) *World Regions in Global Context: Peoples, Places, and Regions, 1st Edition,* p. 5–11. They have been modified slightly to provide you with examples relevant from the 3rd Edition.

referred to as *mental maps*) are psychological representations of locations that are made up from people's individual ideas and impressions of these locations. These representations can be based on people's direct experiences, on written or visual representations of actual locations, on hearsay, on people's imagination, or on a combination of these sources. Location in these cognitive images is fluid, depending on people's changing information and perceptions of the principal landmarks in their environment. Some things, indeed, may not be located in a person's cognitive image at all!

Now turn to the Appendix that begins on page A-1 (Appendix) in the back of the Marston textbook and read the entire section. When you complete that section, you are ready to answer the questions that complete this particular chapter. Enjoy and good luck.

Testing Your Map Knowledge

Fill in the blanks and answer the following questions based on your reading.

1. Lines of _____ run east to west and measure north to south, while lines of _____ run north to south and measure east to west.

2. 0^0 longitude is called the _____, and 0^0 latitude is called the _____.

3. When locating a place solely based on its name, we refer to that as its _____ location. On the other hand, when a place is given by its coordinates in latitudinal and longitudinal space, we refer to that as its _____ location.

4. A _____ is an effective tool for measuring the exact location of particular place through the use of satellite technology.

5. The relative location of a place is fixed in terms of its _____ and _____.

6. Site refers to _____, while situation is used to refer to
_____.

7. Our psychological representations of the world are often called _____, which are also called mental maps.

8. The Peter's projection is a(n) _____ projection that distorts _____.

9. _____ is a term that refers to the collection of information about parts of the earth's surface by means of aerial photography or satellite imagery.

10. Which of the following is the smallest scale?

    (a) 1/10,000   (b) 1/100,000   (c) 1/1,000,000   (d) 1/1,000

11. Using Map A.8, answer the following questions.

    (i). Which of following countries has a natural population increase above 3?

        (a) Iran   (b) China   (c) India   (d) Japan

    (ii). Calling someone in which of the following cities costs the most from the United States?

        (a) Stockholm, Sweden   (b) Sydney, Australia

        (c) Tokyo, Japan   (d) Seoul, South Korea

12. The most time-consuming aspect of GIS work is called _____, which is when the geographer enters data into the system.

13. _____ utilizes census and commercial data to conduct studies, such as market research.

CHAPTER 1

A World of Regions

Learning Objectives

After reading and studying this chapter, you should be able to

1. distinguish regional geography from human or physical geography;
2. comprehend the concepts geographers use to explain interdependence between global regions and between global processes and particular places;
3. have a working knowledge of the processes of globalization and the impact of those processes on place;
4. compare and contrast the different ways in which geographers have examined the relationship between humans and the environment;
5. distinguish the differing processes that have led to the current physical formation of the earth as a system;
6. comprehend the difference between weather and climate and how these two processes impact the organization of various ecosystems;
7. understand the current political and economic configuration of the world-system in a historical perspective;
8. evaluate the changing geography of world regions in relation to the complex relationships between the core and the periphery of the world capitalist system; and
9. comprehend how "development" has been theorized and implemented in different parts of the world.

Vocabulary Review

1. Geographers often use the phrase _____ to designate a large-sized territory made up of a number of different _____ that have similar attributes.

2. Economies of scale refer to the process of _____.

3. When groups of people exhibit similar attributes, such as religious identities, within state boundaries we can refer to this as the process of _____.

4. A cultural ecologist studies the _____.

5. A farmer in Tanzania is working in the _____ sector of the economy, while a woman working in a factory making sneakers in Indonesia is working in the_____ sector of the economy.

6. The Columbian Exchange refers to _____
_____.

7. The _____ is a term that refers to the process by which high birth and death rates are replaced by low birth and death rates.

8. _____ refers to a country's total value of all materials, foodstuffs, goods, and services produced by a country. _____, on the other hand, includes all of the above as well as the value of income from abroad.

9. When one country asserts that a minority population living outside its formal borders belongs to it, that is called _____.

10. A _____ is networks of labor and production processes that originate in extraction or production of raw materials and end in the consumption of finished products.

11. When the location of interrelated technologies, such as transportation technologies, are clustered together we call this a _____ .

12. _____ refers to the theory of how Earth's outer layer has been structured over time.

13. _____ refers to daily atmospheric conditions, while _____ refers to the average of atmospheric conditions over an extended period of time.

14. Where air rises vertically due to intense solar heating near the equator we see the development of the _____ , often abbreviated as _____ .

15. The name assigned to the classification of ecosystems into major types of vegetation and climate is a _____ . A person interested in studying the different biomes and/or ecosystems would probably study_____ .

16. _____ is a term often used to describe the process by which increasing levels of carbon dioxide and other gases are trapping heat in the earth's atmosphere. This process is sometimes called the _____ .

17. In contrast to an ecosystem, a _____ is defined as the increasing political and economic interdependence between countries at a global scale.

18. The core of the current world system could be found in _____ in 1800 and eventually spread to encompass three key cores in by 2000, which include _____ , _____ , and _____ .

19. World empires are important because they facilitated the large-scale distribution of ideas and goods, often through the twin processes of _____ and _____ .

20. When a previously homogenized group becomes spatially dispersed, we often refer to that group as a _____ . On rare occasions one homogeneous group will occupy the boundary of a nation-state, but in most cases states are multinational, and _____ , the feeling of belonging to a nation, must be actively constructed.

21. _____ is the term one would use to describe the domination of one set of peoples over the world economy.

22. A hegemony will often impose a new _____ on its colonial holdings through the development of geographically specific patterns of specialized production.

23. Even though most countries today have been decolonized, one can still find many instances of what is now referred to as _____ , through which powerful states in the core still maintain influence over the economy and politics of peripheral countries.

24. Development theory is _____ .

25. Global neoliberal economic policies are often carried out by two macro-level international organizations called the _____ and the _____ , which is more commonly known by its acronym _____ .

26. _____ is supportive of decreases in government spending as well as the privatization of all government holdings.

27. A _____ or TNC is an organization that has investments that transect international borders and involve the labor and resources of numerous countries.

Activities & Problems

1. According to Figure 1.12, which world regions have more than 25% of their population currently using the Internet?

2. Provide a few hypotheses as to why most of Africa remains without Internet connections.

_____

_____

_____

3. Referring to Figure 1.17 in your text identify the types of boundary between the following plates.

    (a) North American Plate and the Juan de Fuca Plate _____

    (b) South American Plate and the Nazca Plate _____

    (c) Philippine Plate and the Eurasian Plate _____

    (d) Indo-Austrian Plate and the Eurasian Plate _____

    (e) The Pacific Plate and the Philippine Plate _____

4. Identify the following features on the figure below: (a) tropic of Capricorn; (b) tropic of Cancer; (c) Hadley cells; (d) ITCZ; (e) trade winds; (f) westerlies; (g) southeast trade winds; and (h) northeast trade winds.

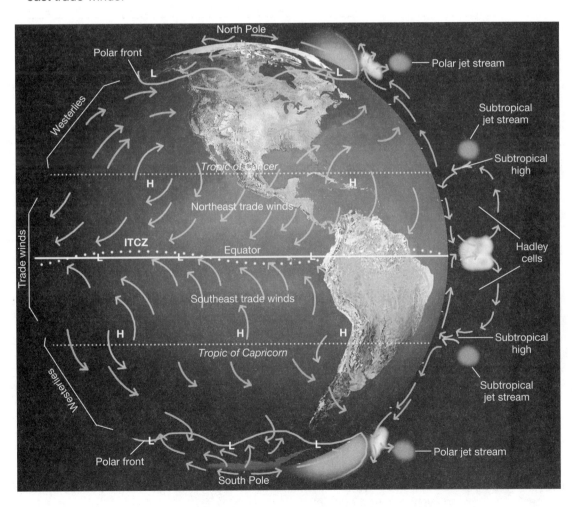

5. Based on the figure in question 4. explain why the ITCZ is located near the equator.

_____

_____

_____

6. Based on Figure 1.22 in your textbook name three central African countries that have a tropical climate: _____ , _____ , and _____ .

7. Using the same figure as in question 4, name three countries in Southeast Asia that have mesothermal climatic regions: _____ , _____ , and _____ .

8. Based on Figure 1.30 in your textbook compare the spatial extent of France in 1714 and in 1914. Do the same for Spain. What is the major difference between the two?

_____

_____

_____

_____

_____

_____

Review Exam

1. What does a regional geographer study? How does that differ from the study of human or physical geography?

_____

_____

_____

_____

_____

2. Discuss the ways in which regions and places exert an influence on people's lives.

_____

_____

_____

_____

_____

3. Briefly discuss how the process of globalization has fostered interdependence between places. Use an example to support your argument.

_____

_____

_____

_____

_____

_____

_____

4. According to the World Commission on the Environment and Development, what is sustainable development?

_____

_____

_____

_____

_____

_____

_____

5. What is the difference between Gross National Product (GNP) and Gross Domestic Product (GDP)?

_____

_____

_____

_____

_____

_____

6. What accounts for the demographic transition?

_____

_____

_____

_____

_____

_____

_____

7. What happens when a country gets caught in a "demographic trap"?

_____

_____

8. What is meant by the term "sense of place"? How is sense of place different for "insiders" and "outsiders"?

_____

_____

_____

_____

_____

9. Briefly describe the five technological phases of the world economy since the initiation of the Industrial Revolution.

_____

_____

_____

_____

_____

_____

10. What is meant by the term "spatial justice"? How might a geographer study spatial justice in relation to gender relations globally? Use examples to support your argument.

_____

_____

_____

_____

_____

_____

11. Briefly describe Alfred Wegner's theory of continental drift.

_____

_____

_____

_____

12. What is meant by the term biodiversity? Why do we have such diverse bioregions today?

_____

_____

_____

_____

_____

_____

13. Describe the changing geography of core, semiperiphery, periphery, and external areas of the world system between 1800 and 2000.

_____

_____

_____

_____

_____

14. What is meant by the term "triadic core"?

_____

_____

15. What are the five main factors that led to the consolidation of the European core beginning in the 15th century?

_____

_____

_____

_____

_____

16. Compare and contrast the terms "nation" and "nation–state." What role does "nationalism" play in the formation of these particular entities?

_____

_____

_____

17. What role did the development of internal infrastructures play in the transformation of core economies? What helped facilitate these changes?

_____

_____

_____

18. What is meant by the term "international division of labor" in the context of colonization?

_____

_____

19. Briefly describe the changing geography of Indian lands in the United States between 1790 and 2000.

_____

_____

20. What impact did colonization have on the transfer of crops around the world? Using the potato as an example, briefly describe the historical geography of that crop's movement across the globe.

_____

_____

_____

_____

_____

_____

21. How did Rostow theorize development? What did his detractors, particularly dependency and feminist theorists, have to say about Rostow's theory?

_____

_____

_____

_____

_____

_____

22. Briefly describe the spatial distribution of climatic zones at the global scale in relation to the major world regions covered in the text.

_____

_____

_____

_____

_____

_____

23. Describe the relationships between climatic zones and world ecosystems at a global scale.

_____

_____

_____

_____

_____

24. The interdependence between the Old (Europe) and New Americas) Worlds is often referred to as ?

    (a) Old-New World Connection

    (b) Globalization

    (c) The Columbian Exchange

    (d) Regionalization

25. The generic term for the process by which one ethnic group seeks to create autonomy for themselves within a national context is often called _____?

    (a) irredentism   (b) regionalism   (c) sectionalism   (d) nationalism

26. When one state exerts political and legal domination over another territory, this is called:

    (a) economies of scale   (b) imperialism

    (c) jingoism   (d) colonialism

27. Which of the following terms is used to describe a region that demonstrates a high degree of homogeneity?

    (a) formal region   (b) functional region

    (c) regionalism   (d) sectionalism

28. When a collection of national states formally come together to address a common goal, we call the resultant structure a _____.

    (a) world region   (b) supranational organization

    (c) global community

29. The economic success of the entertainment industry in the United States has promoted the notion of a growing global culture based on _____.

    (a) Europeanization   (b) McWorldization   (c) Americanization

    (d) colonization

30. When one country asserts a historic and cultural right over a minority living outside of its current national borders, it is making a(n) _____ claim.

    (a) sovereignty   (b) colonialist   (c) imperialist   (d) irredentist

31. Someone working in the area of knowledge and information technology is working in which sector of the economy?

    (a) primary   (b) secondary   (c) tertiary   (d) quaternary

32. By the year 2000 the core of the global economy consisted of all of the following except?

    (a) Canada   (b) Japan   (c) United Kingdom   (d) Russia

33. _____ is the process by which trade winds pick up water over the ocean and then drop that water as rain at the top of coastal mountains.

    (a) biogeography   (b) orographic lifting

    (c) biome development   (d) global warming

34. Which of the following transnational organizations provides loans to governments throughout the world?

    (a) IMF   (b) TNC   (c) World Bank   (d) GATT

35. Regions of the world known for being in a completely disadvantaged and dependent position are called _____.

    (a) core regions   (b) peripheral regions

    (c) semiperipheral regions   (d) external regions

36. The term "diaspora" refers to which of the following?

    (a) the unification of a formally homogeneous group
    (b) the unification of formally heterogeneous groups into a unified group
    (c) the spatial dispersion of a previously homogeneous group
    (d) the spatial dispersion of a number of heterogeneous groups

37. During the _____ incoming solar energy is greatest in the Northern Hemisphere?

    (a) vernal equinox   (b) winter solstice   (c) autumnal equinox   (d) summer solstice

38. Which of the following is not a key factor in the process of "globalization"?

    (a) a new international division of labor

    (b) the internationalization of finance

    (c) transnational corporations

    (d) mercantalism

Mapping Global Life Expectancy[1]

1. In this exercise, you are now going to examine average life expectancy globally. Begin by taking a clean copy of the map provided and drawing boundaries around the 18 world regions listed in the "2006 World Population Data Sheet" (WPDS) located at the back of this guide. (NOTE: Aside from "North America" and "Oceania" the other 16 regions are listed as subregions of larger continental spaces, such as Africa, Europe, Latin America and the Caribbean, etc.)

2. Now, looking at the WPDS find the column labeled "Life Expectancy at Birth (years),Total." Using a table, list the world regions and their corresponding life expectancy figure from greatest to least. Rank these expectancies from greatest to least, placing a number 1 next to the greatest life expectancy and an 18 next to the lowest.

---

[1]Answers to Map Exercises will vary and therefore no answers are given in the corresponding Answer Section below.

3. On your map of world regions, label each region with its corresponding number (and life expectancy mean in brackets). Answer the following questions:

   (a) Which world regions have the lowest life expectancy? Which regions have the highest?

   _____

   _____

   _____

   _____

   (b) Do you notice a spatial pattern in the statistics? Are there any distinctive lines that can be drawn, globally, between the top five regions and the five lowest regions? If so, what are they?

   _____

   _____

   _____

   _____

   _____

   _____

   _____

   _____

   (c) Going back to your original statistics, which specific countries seem to have the lowest life expectancy in relation to their regions and to the world (you can calculate a world mean using the averages of all the world regions)? Why do you think these countries stand out? Explain.

   _____

   _____

   _____

   _____

   _____

   _____

   _____

   _____

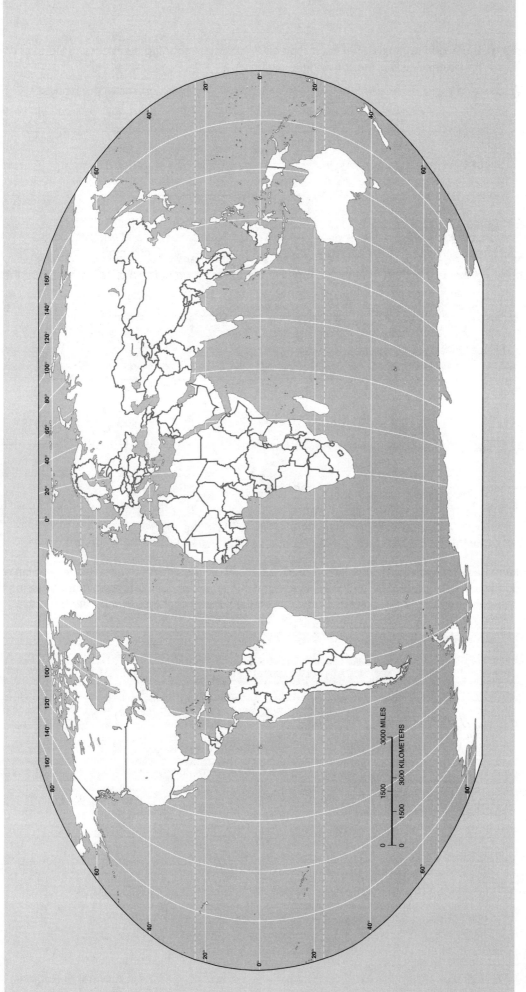

MAP 1.1

CHAPTER 2

Europe

Learning Objectives

After reading and studying this chapter, you should be able to

1. understand the complex physical geography of Europe as well as its environmental history;
2. have a working knowledge of the relationship between industrialization and imperialism;
3. distinguish the distinct cultural patterns that are part and parcel of the complex population geography of Europe;
4. compare and contrast the development policies of "western" and "eastern" Europe;
5. examine the distinct patterns of population and migration that distinguish this region;
6. comprehend the dynamics of regional integration and difference in Europe;
7. describe the geography of Europe's core regions and key cities in relation to more peripheral areas in the region; and
8. understand the importance of political and economic integration in the region and the pace at which that integration is occurring.

Vocabulary Review

1. The four main physiogeographic regions of Europe are _____, _____, _____, and _____.

2. Where seas have risen to meet glaciated landscapes one can often find _____.

3. In the Netherlands the local population perfected the practice of draining lands through the use of windmill pumps, drainage ditches, and canals. The results of their efforts were flat, fertile, stone-free land, or what is also called a _____ landscape.

4. Industrialization in Europe had profound impacts, including the production of _____, which resulted from heavy doses of atmospheric pollution related to factory production.

5. Eastern Europe was often considered to be a _____ between the Soviet Union and western Europe during the Cold War period. The countries caught between the two powerful blocks were often thought of as _____ states of the Soviet Union.

6. Despite a socialist or communist rhetoric, the states of eastern Europe and the Soviet Union developed into _____ economies, which were highly bureaucratic and state centered.

7. The process by which larger states are broken up into smaller states, particularly around ethnic groups, is called _____.

8. In some cases ethnic minorities do not break away from a country but instead remain in _____ within another state.

9. In some cases extreme nationalism, which is sometimes referred to as chauvinism, has led to the extreme practice of _____, whereby one ethnic group is forcibly evicted from a particular place.

10. Immigration creates both positive and negative results. In some cases, immigration of ethnic minorities into countries in Europe has led to a rise in _____, hate or fear of a minority group.

11. During the early part of the 20th century, industrial employment was scaled back in places such as northern England, south Wales, and central Scotland. This process is called _____, and it often is accompanied by _____, which includes the selling off of assets such as factories or equipment.

12. The _____ is the modern core region of Europe that stretches from London to Paris to Berlin.

13. The development of a centralized, autocratic government, which values nation and race over individuals, is often defined by the term _____.

14. In London most of the older inner-city housing has disappeared, replaced by _____ projects.

15. _____ is defined as a way of life based on the herding and grazing of animals. More often than not regions where such practices are present have relatively low-density populations.

16. As one moves north in Europe, the vegetation patterns become consumed by gray lichens and dwarf willows and birches, and forests cover over half the land. This area is best representative of a _____ biome.

17. On the Danubian Plain, particularly in the Hortobágy area, one would find semiarid, treeless grassland, also representative of a _____ biome.

18. In the Mediterranean region of Europe successive empires carved out large estates, known as _____, where peasants were often forced to work. Land that was not part of these larger estates was sometimes parceled off into smaller properties by peasant farmers who called them _____.

19. Intermediary cities, or _____, were seaport towns that served as trading and trans-shipment points throughout Europe.

Activities & Problems

1. On the following map identify the four major physiographic regions by name: (a) northwestern uplands; (b) north European lowlands; (c) central plateaus; and (d) Alpine system

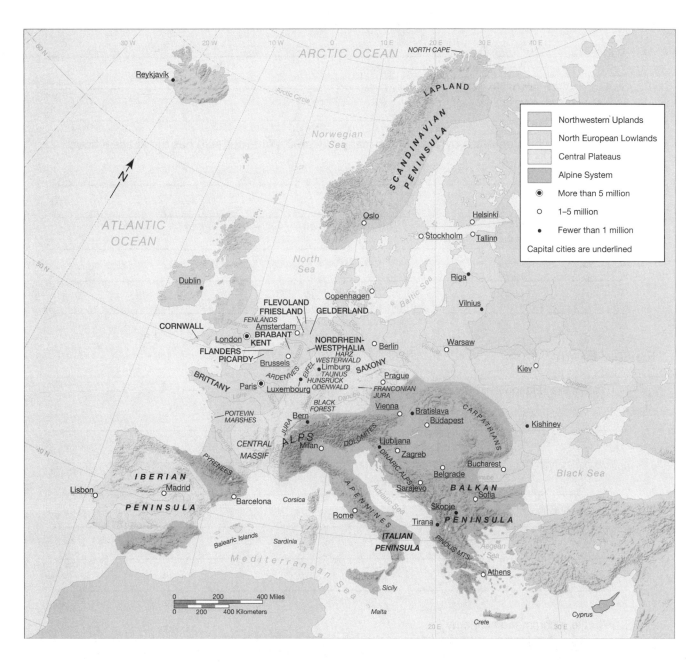

2. Between 1945 and 1973 a significant amount of the migrant population came from outside Europe. According to your reading of Figure 2.18 in your text, what non-European population was the largest migrant group into Germany? _____ Into France? _____

3. Between 1945 and 1973, there was a significant amount of labor migration between countries in Europe (see Figure 2.18 in your text). Internal migration within the region, however, was not even, and some countries had much higher numbers of out-migrants. Name the seven countries within Europe that witnessed significant out-migration in this particular period: _____, _____, _____, _____, _____, _____, and _____.

4. Give two reasons why the internal migration patterns of Europe between 1945 and 1973 look the way they do.

_____

_____

_____

_____

_____

5. Explain why the population pyramid of Germany (Figure 2.17 in your text) has the shape it does.

_____

_____

_____

_____

_____

Review Exam

1. Explain the relationship between trade and the "Age of Discovery."

_____

_____

_____

_____

_____

_____

2. What is a command economy?

_____

_____

_____

_____

_____

_____

3. What impact has the European Union had on agriculture in Europe?

_____

_____

_____

_____

_____

_____

4. What four advantages have allowed Europe's "Golden Triangle" to thrive?

_____

_____

_____

_____

_____

_____

5. Describe the characteristics of the "Southern Crescent." Where is it and why has it become a place of relative prosperity?

_____

_____

_____

_____

_____

6. Briefly describe the diversity of Europe's Muslim population, considering why there has been an increase in Muslim in-migration to Europe since World War II.

_____

_____

_____

_____

_____

7. Which language group is most commonly used on the Iberian Peninsula?

   (a) Romance   (b) Germanic   (c) Celtic   (d) Baltic

8. A person living in Sicily would be living in which physiogeographic region?

   (a) northwestern upland

   (b) north European lowland

   (c) central plateaus

   (d) Alpine system

9. According to Figure 3.21, which of the following regions are eligible for aid from the European Union's regional development program?

   (a) southern portion of the United Kingdom

   (b) southern portion of Spain

   (c) Switzerland

   (d) most of Western Germany

10. Which of the following areas in Europe still had not been impacted by industrialization by 1914?

    (a) northern Ireland

    (b) western France

    (c) northern Italy

    (d) southern Italy

11. Portuguese workers have tended to seek employment in which country?

    (a) Spain   (b) United Kingdom   (c) France   (d) Germany

12. The _____ movement was based on the conviction that all of nature, as well as human beings and societies, could be understood as a rational system.

    (a) poststructuralism   (b) enlightenment   (c) communism   (d) postmodernism

13. Which of the following is not true about agricultural production in the Golden Triangle of Europe?

    (a) it is highly intensive

    (b) it is geared toward the rural hinterlands

    (c) it is focused on fresh dairy, vegetables, and flowers

    (d) it is used to supply the city

14. The growth of the Southern Crescent, a secondary core region in Europe, has resulted from which of the following?

    (a) the decentralization of industry from northwest Europe

    (b) the influx of capital from Middle Eastern oil companies

    (c) the expensive labor in the southern part of the region

    (d) the slow rate of urbanization in the region

15. Which of the following best characterizes nordic Europe?

    (a) Most of the important towns are inland, located far from the sea.

    (b) Over 50% of the land is used for agriculture across the entire region.

    (c) The orientation of people in this region is toward the interior uplands.

    (d) The orientation of people in this region is toward the sea.

8. Offer several hypotheses that might explain the pattern (cite key points in the textbook that help support your hypotheses).

_____

_____

_____

_____

_____

_____

_____

_____

_____

_____

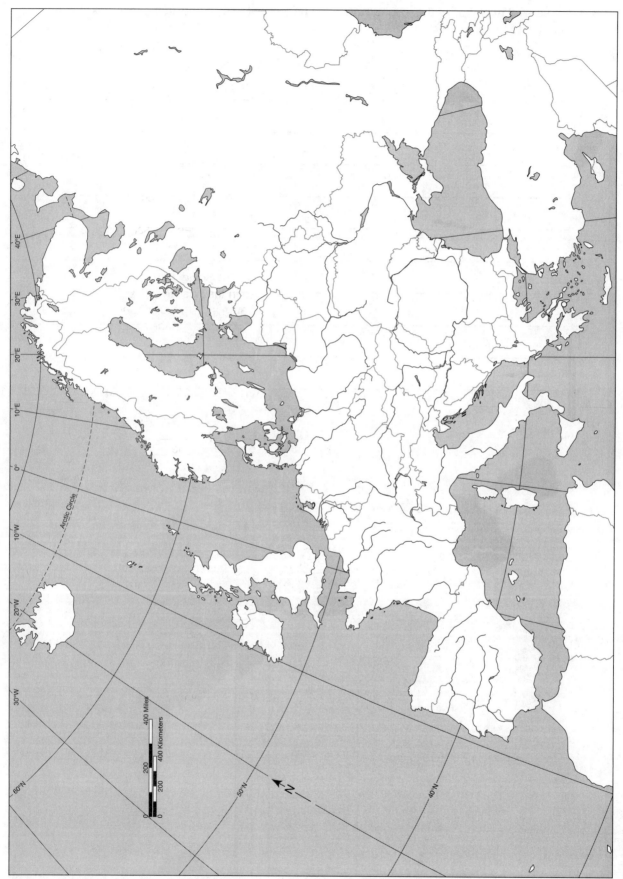

Map 2.3

CHAPTER 3

The Russian Federation, Central Asia, and the Transcaucasus

Learning Objectives

After reading and studying this chapter, you should be able to

1. understand the unique physical geography of this region;
2. have a working knowledge of the important role of primary commodities, such as fur, to the historical expansion of Russia and Central Asia;
3. comprehend the history of the Soviet State and the relationship between Russia and the former Soviet Republics;
4. compare and contrast subregional development trajectories;
5. explain the role that ethnic diversity played in the breakup of the Soviet Union;
6. understand the importance that organized crime has played in the post-Soviet states;
7. comprehend the historical changes in the population geography of the region; and
8. explain the problems related to political, economic, and social change in the region.

Vocabulary Review

1. In the _____ region it is unlikely that there would be any agricultural production or forestry.

2. The largest physiogeographic zone in the region is the _____, which refers to an area that is made up of large tracts of spruce, fir, and pine trees. The _____ region is a flatter landscape devoid of trees and dominated by tall grasses.

3. Black earth, known by its technical name _____, can be found throughout the _____ region and has a high level of fertility.

4. In contrast to the Czarist _____ state, which was highly centralized, the Leninist model for the Union of Soviet Socialist Republics (USSR) was a _____ state in which power was to be devolved to the local republic level.

5. In the new Soviet state the central unit of political, economic, and social organization was the _____, a community that was constituted by a 300- to 800-meter radius and was home to 8,000 to 12,000 people.

6. In the post-Soviet era, despite attempts to liberalize the state, _____ organizations, such as nongovernmental organizations, business organizations, and pressure groups, have not emerged.

7. In the northernmost part of this world region the ground is constantly frozen, a condition known as _____.

8. Under _____ basic services, such as housing, health care, and education, were provided to the citizens of the Soviet Union at little to no cost.

9. During the Soviet period large tracts of the eastern steppe region were cultivated and wheat was grown in large quantities. Lack of rainfall in the region, however, demanded that farmers turn to _____ techniques and irrigation, which included the constant weeding of fields and leaving the stubble of plants to trap snow in the winter.

10. Uzbekistan has suffered from large-scale agricultural production of cotton. In particular, _____, which is caused when water evaporates and leaves salt behind on top of the soil, has slowly diminished the productive capabilities of cotton farmers in this now-independent state.

11. Because the Soviet industrial economy was not based on free markets, the government organized _____, a set of regionally based production facilities to maximize productivity.

12. Central Asia was tied into the world economy through the 11th century because the _____ ran directly through, and between, the small agricultural oasis communities of this subregion.

13. In the post-Soviet period this region has been marked by several _____ and _____ claims, the latter of which includes Chechnya's attempt for political independence and the former of which includes the claims of Tajikistan and Kazakhstan over territories currently outside their own political borders.

14. The _____ was a metaphor used to describe the territorial extent of Soviet control in eastern Europe after 1948 and prior to the breakup of the Soviet Union.

15. The _____ form a physiogeographic boundary between the Russian and West Siberian plains.

16. The _____ is an inland sea that is choked off by the Bosporus and Dardanelles Straits as well as the Sea of Marmara.

17. The political hierarchy of the Soviet State included not just 15 Soviet Socialist Republics but also 20 autonomous Soviet Socialist Republics, 8 autonomous _____ or regions, and 10 autonomous _____ or areas.

18. The _____, or state bureaucrats, were key actors in the organization and development of the Soviet State.

19. In an attempt to combat the economic upheaval caused by the breakup of the Soviet Union, the Russian Federation led a coalition of new states in the formation of the _____, or CIS, which was a forum to discuss economic and political problems in the region.

20. The _____ is a region made up of former Soviet Republics with large numbers of ethnic Russians in which the Russian Federation still wants to assert its ongoing influence.

Activities & Problems

1. Using Figure 3.18 in your text, answer the following questions:

   (a) In what century did the Russian State expand out to the Caspian Sea? _____

   (b) In what century did the Russian State expand out to the Black Sea? _____

   (c) In what century did the Russian State expand out of the Sea of Okhotsk?

   _____

   (d) In what century did the Russian State acquire the land to establish the city of St. Petersburg?

   _____

2. Using Figure 3.19 in your text, answer the following questions:

   (a) What three territories were incorporated into the USSR in 1940? _____, _____, _____

   (b) What three eastern European countries were incorporated into the Soviet block in 1947? _____, _____, _____

   (c) When did the Soviet Union occupy the eastern part of Austria? _____

   (d) During what time period was Albania part of the Soviet bloc? _____

   (e) What three countries lost part of their territory to the Soviet Union between 1940 and 1947?

   _____, _____, _____

3. Compare and contrast the major differences in the physical geography of the West Siberian Plain, the Central Siberian Plain, and the Russian Plain.

_____

_____

_____

_____

_____

_____

4. Compare and contrast the different vegetation patterns found in tundra, taiga, and steppe landscapes of this region.

_____

_____

_____

_____

_____

_____

5. Using Figure 3.27 describe the different geographic patterns of agricultural production under collectivization and decollectivization.

_____

_____

_____

_____

_____

_____

Review Exam

1. Describe the physiographic regions of the Russian Federation, Central Asia, and the Transcaucaus.

_____

_____

_____

_____

_____

2. Under Stalin how did Soviet industrial policy change? What were the impacts of that change?

_____

_____

_____

_____

_____

_____

3. Define the terms perestroiyka and glasnost.

_____

_____

_____

_____

_____

_____

4. What impact did decollectivization have on the layout of villages in the former Soviet Union?

_____

_____

_____

_____

_____

_____

5. What were some of the legacies of Soviet rule that facilitated the spread of crime and corruption in the post-Soviet state?

_____

_____

_____

_____

_____

_____

_____

6. Which of the following is true about the Central Region?

   (a) It is highly urbanized.

   (b) It is highly rural.

   (c) Moscow is the third largest city in the region.

   (d) About 75% of the region is forested.

7. Which of the following is not true about post-Soviet Moscow?

   (a) The transition has been smooth with little conflict.

   (b) The housing market has slowly risen to meet the demand of the growing population.

   (c) The construction of new homes has grown uncontrolled in the forest protection belt that surrounds the city.

   (d) There has been little appearance of Western-style stores or shops.

8. An area of sparse vegetation with no agricultural or forestry production capabilities is called a

   _____ .

   (a) steppe   (b) chernozem   (c) taiga   (d) tundra

9. Which of the following is not the name of a political unit in the Russian Federation?

   (a) Okrug   (b) Kray   (c) State   (d) Republic

10. In 1994, Russian troops invaded which of the following republics, kicking off seven years of warfare?

   (a) Georgia   (b) Chechenya   (c) Dagestan   (d) Adygeya

11. Which of the following was one of the major reasons for the early expansion of the Russian Empire out of Muscovy and into Siberia?

   (a) desire for oil reserves

   (b) desire for an outlet to the Pacific Ocean

   (c) desire to obtain more forest resources, such as furs

   (d) desire to trade more effectively with the Chinese

12. Which of the following was not a consequence of the Soviet system?

  (a) regional equality

  (b) environmental degradation

  (c) diffusion of ethnic Russians throughout the region

  (d) the suppression of ethnic-based nationalism

13. What is the zone of boreal coniferous forest called?

  (a) steppe   (b) taiga   (c) tundra   (d) chernozem

14. Which of the following is true of the Aral Sea?

  (a) It has shrunk by more than 40% of its surface area.

  (b) It has increased by more than 40% of its surface area.

  (c) Its sea level has dropped by more than 33 meters.

  (d) Its sea level has risen by more than 33 meters.

15. This particular region has been one of the historical cores of Soviet industrialization and is known for its high levels of coal, oil, and natural gas.

  (a) the Urals   (b) the Volga region   (c) the Steppes   (d) the Central Siberian Tiaga

For each of the following statements, write T for "True" or F for "False" on the space provided.

16. There has been no privatization of forest resources in the post-Soviet period in the Siberian taiga region. _____

17. Many of the former republics in the central Asian deserts suffer today from salinization. _____

18. The breakup of the Soviet Union in 1989 led to the suppression of ethnic uprisings in a number of federation members in the 1990s. _____

19. By the year 2000 all of the state-owned businesses in the Russian Federation were sold to private investors. _____

20. Over the course of the 1990s, agricultural output increased at a tremendously rapid rate throughout the Russian Federation. _____

21. The Russian Federation has begun to take on the nuclear waste of other countries as a means to increase foreign reserves. _____

22. Between 1950 and 1955 economic output grew by nearly 10% per year in the Soviet Union. _____

23. The movement of Russians into the central Siberian taiga region in the 16th century did little to disrupt the lives of the local Tungus and Yakuts peoples. _____

24. The Urals were an economically unimportant and marginal component of the Russian Empire in the 18th and 19th centuries. _____

25. Moscow has done little to alter its urban planning philosophy since the fall of the Soviet state. _____

7. How and why might this population map change over the next 10 years? Explain.

_____

_____

_____

_____

_____

_____

_____

_____

_____

_____

_____

_____

_____

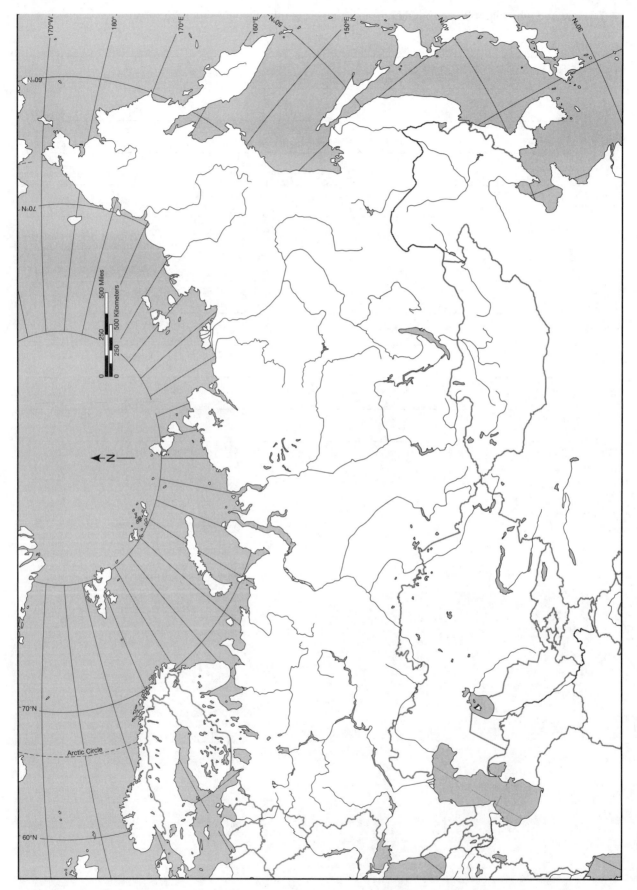

Map 3.3

CHAPTER 4

Middle East and North Africa

Learning Objectives

After reading and studying this chapter, you should be able to

1. have a working knowledge of the region's landforms and landscapes;
2. understand the historical and geographic development of the early empires of this region as a source of domesticated plants and world religions;
3. comprehend the basic tenets of Islam as a religious philosophy and practice and the growth of this religion in relation to Christianity and Judaism;
4. compare and contrast the differential impacts of empire building and colonization on the region;
5. evaluate the complex social and spatial systems that mediate class, gender, and ethnicity in the region;
6. understand the importance of natural resource development—water, oil, agriculture, etc.—in the region;
7. describe the different political tensions in the region and their concurrent political geographies; and
8. have a working knowledge of the distinctive urban and rural geographies within the various subregions of the Middle East and North Africa.

Vocabulary Review

1. The greatest defining climatic feature of the Middle Eastern and North African region is _____ , in that the climate lacks enough moisture to support trees or woody plants.

2. A(n) _____ can "rise out of the desert" because of both a low water table and land erosion that exposes the water table to the surface.

3. In parts of Turkey farmers can practice "dry farming," which means they farm _____ .

4. To combat the reduction in forest coverage, some countries are engaging in reforestation, or _____ programs.

5. The _____ were the largest indigenous ruling group in the region prior to colonization of the region by Europeans.

6. The _____ , instituted by the British, established a homeland for Jews in _____ in 1917.

7. Economic activities that take place outside the "official records" of the government are called _____ , and are often an important part of daily life for many people in the region.

8. The 1930s, which was the height of European colonization in parts of the region, was also a time when certain states, such as Turkey and Egypt, began a policy of _____ , whereby domestic producers provided goods that were previously provided by foreigners. In the post-WWII period, on the other hand, national policies became even stricter as a number of countries followed a policy of _____ of all industry, turning private enterprises into state enterprises.

9. Islam is a term that means _____ , whereas the term _____ refers to a person "whose duty is obedience and submission to the will of God."

10. _____ is often mistaken for fundamentalism, which refers to a strict adherence to the religious doctrine. But the former actually refers to _____ .

11. The term _____ is often interpreted incorrectly as "holy war," while in reality the term is defined as "peaceful struggle" both internally and externally.

12. While the term _____ is often applied to biological relationships, it can also be extended to include ties among broader social groups, such as neighbor and friends.

13. Pastoralism is best defined as _____. In many cases pastoralists practice _____, moving herds based on seasonal changes.

14. In some cases, Muslim women wear the _____, a full body covering, while in other instances women wear only a _____ covering their head.

15. Throughout the Gulf States, which are all rich in oil reserves, a large proportion of the labor force, almost 75%, are _____, brought in from other countries to work in oil production.

16. The _____ is the sacred pilgrimage to Mecca that each financially and physically able Muslim man should make over the course of his life.

17. The Palestinians, in an act of defiance against the Israeli state, began a(n) _____ or uprising in the late 1980s to protest for an independent Palestine.

18. _____, or revenue that results from the sale of oil, are often reinvested into the larger world system of banking and investment.

19. _____ was a late 19th century movement that pushed for the development of a legal homeland for Jews.

20. Throughout Sudan, Turkey, and Iran there is a new and emergent group of individuals labeled as _____ because they have refugee status within their own country.

Activities & Problems

1. Based on Figure 4.18, name the colonial power that occupied the following countries:

   (a) Morocco _____   (d) Egypt _____

   (b) Kuwait _____   (e) Tunisia _____

   (c) Libya _____   (f) Anglo-Egyptian Sudan _____

2. Based on Figure 4.24, answer the following questions about restrictions on women's movements.

   (a) In which countries do all of the following apply: women are secluded by some ethnic groups; there exists a compulsory government-mandated dress code; there have been attacks on women for immodest dress; and women need permission from a male relative to get a passport to travel abroad?

   _____

   (b) Which countries do not require a male relative to give permission for a woman to get a passport?

   _____

   (c) Which North African countries have reported attacks on women for immodest dress?

   _____

   (d) In which countries do none of the following apply: women are secluded by some ethnic groups; there exists a compulsory government-mandated dress code; there have been attacks on women for immodest dress; and women need permission from a male relative to get a passport to travel abroad?

   _____

3. Examining Figure 4 27 in your text, explain migration patterns in the region in 1990. When did these patterns begin to develop? What was the impetus for the migration patterns we see today?

_____

_____

_____

_____

_____

_____

4. Based on your reading of Figure 4.33, how did the map of Israel change between 1949, 1967, and 1996? How does the map differ today compared with 1996?

_____

_____

_____

_____

_____

_____

5. As populations continue to increase, there is going to be more and more pressure on the lands of the Middle East and North Africa. In this question, you are going to: (a) calculate projected population densities for the countries of this region; (b) compare those calculations to the 2006 population density; and (c) examine the total percentage of change in population density over the two time periods.

   (a) Turn to the "2006 Population Data Sheet" located at the back of this text. On a separate piece of paper make the following columns in this order (you may want to use a spreadsheet program, such as Excel, for this exercise): (a) country name, (b) area of country, (c) projected population for 2050, (d) population mid-2006, (e) population density 2006, (f) projected population density 2050, (g) total increase in population density.

   (b) To calculate the population density for 2006 (column e) and 2050 (column f), you must do the following: divide the population mid-2006 and the projected population for 2050 by the surface area.

   Example: Israel (2006) = 7.2 million / 8131 square miles = 885.5 people per square mile

   Israel (2050) = 11 million / 8131 square miles = 1352.84 people per square mile

   (c) Now you can compare 2006 and 2050 projected population density values and fill in column g. We do this by calculating the percentage of change over time. This is done by setting up an equation to compare the two values. In the equation, the 2006 population density is taken to be 100%. We then calculate to see how much more dense it will be in 2015.

   Example: Israel = pop density 2006/100 = pop density 2050/x, solve for x.

   Israel = 885.5/100 = 1352.84/x

Israel $= (1382.54 \times 100)/885.5 = x$

Pop density percent change $= 152.77\%$

Population density in 2050 will be 152.77% of the population density in 2006.

(d) Calculate the percentage of change for all the countries.

(e) Which countries are going to have the highest percentage of change in population density by 2050? Why do you think that is the case?

_____

_____

_____

_____

_____

_____

_____

_____

_____

_____

Review Exam

1. Describe the climatic map of North Africa and the Middle East (i.e., think about spatial patterns and spatial differentiation).

_____

_____

_____

_____

_____

2. What impact did plant and animal domestication have on the peoples of the Fertile Crescent, particularly during the transition to food-producing minisystems?

_____

_____

_____

_____

_____

3. Define the term transhumance. How does transhumance differ from sedentary life?

_____

_____

_____

_____

_____

4. What programs are in place to combat the debilitating effects of desertification?

_____

_____

_____

_____

5. Making particular reference to OPEC, describe the ways in which oil in the region can be examined as both an economic and a political commodity?

_____

_____

_____

_____

_____

6. The Red Sea has formed by which of the following processes?

(a) collision/subduction   (b) collision/continental overthrusting

(c) horizontal displacement   (d) seafloor spreading

7. Algiers is located in which of the following climatic regions?

(a) tropical   (b) dry arid/semiarid

(c) midlatitude   (d) highland

8. In the post-WWII period, a number of countries in the region were led by several key economic policies, including which of the following?

(a) nationalization of industry   (b) import substitution   (c) export-led growth

(d) a and b   (e) a and c

9. The sacred pilgrimage to Mecca, which is one of the Five Pillars of Islam, is called

_____ .

(a) jihad   (b) hajj

(c) Islamism   (d) alms giving

10. Which of the following are both causes of the political tension between Iran and Iraq?

    (a) the historical–cultural differences between Arabs and Persians and the desire for both countries to conquer Kuwait

    (b) the historical–cultural similarities between Arabs and Persians and the desire for both countries to conquer Kuwait

    (c) the historical–cultural differences between Arabs and Persians and the political geographic struggle over the border area known as the Shat-al-Arab

    (d) the historical–cultural similarities between Arabs and Persians and the Iranian support of Iraqi Kurdish minorities in Iraq

11. Which of the following countries are a member of OPEC but not a member of the Arab League?

    (a) (a) Iran   (b) (b) Iraq

    (c) (c) Saudi Arabia   (d) (d) Bahrain

12. Which of the following territories was a former Spanish colony?

    (a) (a) Western Sahara   (b) (b) Algeria

    (c) (c) Libya   (d) (d) Egypt

13. Which of the following are causes of outmigration from the region?

    (a) lack of economic opportunity   (b) war   (c) civil unrest

    (d) all of the above   (e) none of the above

14. Islam is an Arabic term that means _____.

    (a) submission   (b) willingness to serve

    (c) desire   (d) hope

15. The Maghreb has NOT been home to which of the following?

    (a) Romans   (b) Vandals

    (c) Turks   (d) Phoenicians

For each of the following statements, write T for "True" or F for "False" on the space provided.

16. Nearly three-quarters of all workers in the region are guest workers._____

17. The dominant form of Islam in Iran is Sunni Islam._____

18. Intifada is a term that means "uprising" and is used to represent the struggle of Palestinians seeking their own state within the occupied territories of Israel._____

19. Yemen remains one of the wealthiest oil-producing states in the Middle East._____

20. The headwaters of the Tigris and Euphrates Rivers are in modern-day Turkey._____

21. A chador is the sacred room in the mosque in which men pray._____

22. The Byzantine Empire expanded into southern Europe and the modern-day country of Spain._____

_____

_____

_____

_____

_____

_____

_____

_____

_____

_____

6. The Palestinians have been forced to migrate more than other peoples in the region. What are some of the factors that have pushed Palestinians to migrate out of Palestine and Israel? In which countries can you find the most Palestinian migrants? What impact do these migrations have on their host countries?

_____

_____

_____

_____

_____

_____

_____

_____

_____

_____

_____

_____

_____

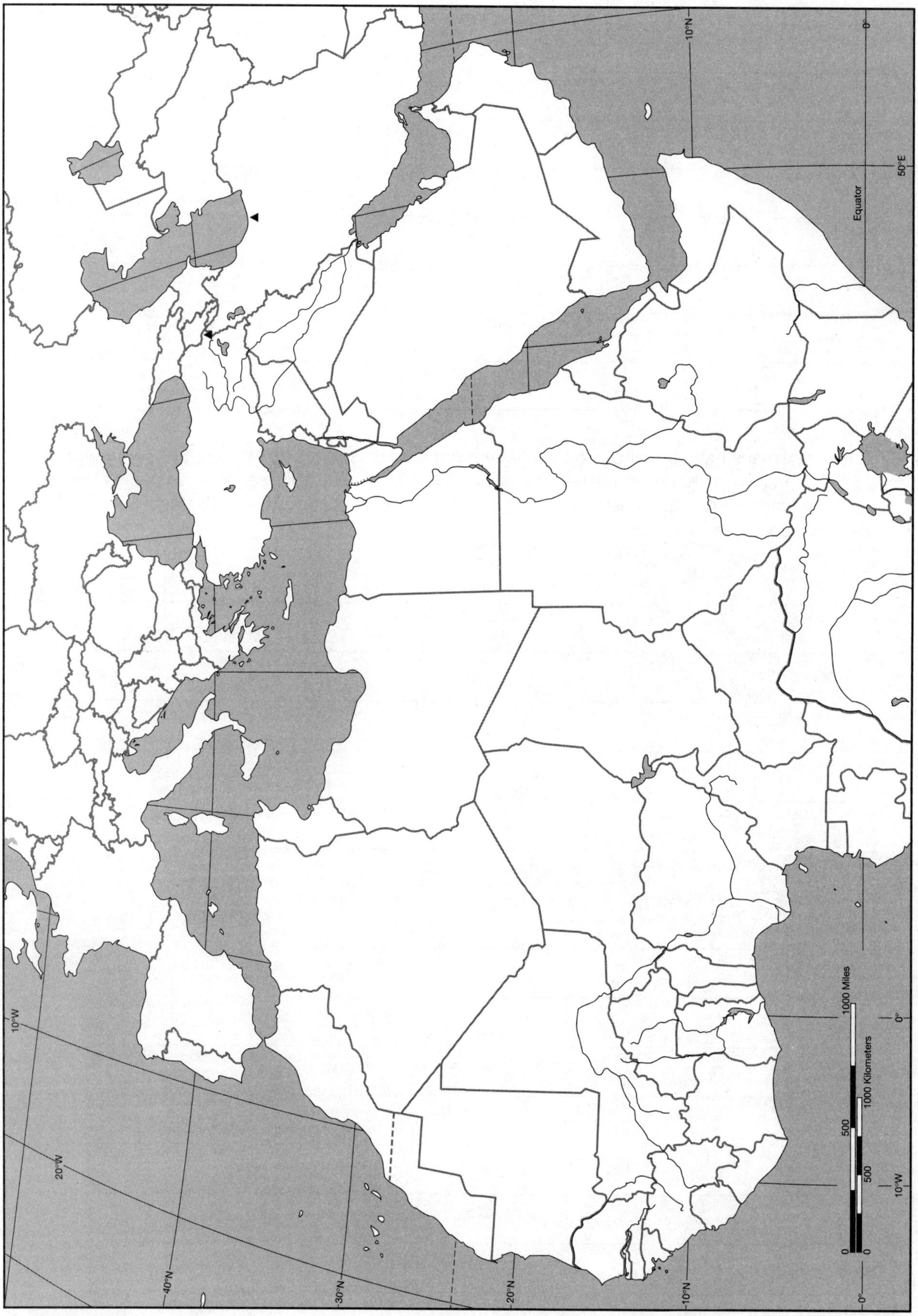
10°N
0°
50°E
Equator
10°W
20°W
40°N
30°N
20°N
10°N
0°
1000 Miles
1000 Kilometers
500
500
0
0
0°
10°W

Map 4.3

CHAPTER 5

Sub-Saharan Africa

Learning Objectives

After reading and studying this chapter, you should be able to

1. have a working knowledge of the relationship between the region's environmental and human geographies;
2. understand the various climatic and tectonic activities and patterns that define this region;
3. comprehend the long and complex history of human civilization in the region;
4. be able to trace the dramatic impact that slavery and European colonialism had on the region;
5. compare and contrast the differential impacts of postcolonialism among the countries of the region;
6. investigate the patterns of culture, language, and religious diversity that are part and parcel of this complex region;
7. examine the important development debates that continue to mark this region's economic geography; and
8. have a working knowledge of the complex urban and rural geographies of this region as well as the interplay between the two.

Vocabulary Review

1. The winds, which are commonly called _____, emerge out of inland Africa carrying large amounts of dust and can cause very hot, dry conditions.

2. Plant _____ was an important part of the rise of civilizations in places such as Ethiopia, where peoples cultivated millet and other important crops thousands of years ago.

3. _____ is one of the ways by which shifting cultivators replenish depleted fields. This process consists of not planting a field for one or more seasons, allowing secondary growth to emerge and reinvigorate the soil.

4. Today, the continent of Africa is made up of numerous countries. These modern-day political boundaries have their origin in Europe's _____, which was the rush to formally colonize all of the Africa initiated after the _____ of 1884–1885, when Otto von Bismark tried to negotiate European holdings on the African continent.

5. European influences on the continent have been numerous. In South Africa, the legacy of European colonization resulted in the establishment of a(n) _____ state under which Blacks and whites were socially and spatially separated.

6. In the late 1950s and 1960s, the South African government put in place its strategy of spatial segregation through the development of _____, which were designed as "tribal reserves."

7. Even though the continent of Africa has born witness to numerous urban development projects, many regions are still marked by _____, the process by which workers respond to the availability of seasonal work.

8. Sub-Saharan Africa is known for some of the highest levels of poverty in the world. Poverty, however, does not impact men and women equally. Women make up over two-thirds of the people living in poverty in the region. This process is known as the _____ and can be found in every region of the world today to different degrees.

9. In order to combat the differential effects of poverty on women, some countries have adopted the _____ approach designed by "Women in Development" to combat these

differences. The goals are to better link women's reproductive and productive roles and understand how the local social and spatial organization of society mediates women's experiences of development.

10. Following the successful implementation of the Grameen Bank in Bangladesh, some countries in Africa have also developed _____ programs, whereby they provide small amounts of credit to families so that they can start their own businesses. These projects are most successful in well-established communities that have high levels of _____, which are networks and relationships that encourage trust and cooperation.

11. The countries of Ethiopia, Somalia, Djibouti, and Eritrea constitute a region known as the _____.

12. _____ is only one form of shifting agriculture, whereby people clear land through burning to make room for domesticated staple crops.

13. _____, a modified version of Dutch, is a language that developed in South Africa among settlers who entered the region almost 400 years ago.

14. The _____ was the term used to define the leg of the slave trade from Africa to the American colonies, in both South and North America.

15. Under British colonial rule, local leaders collected a _____ tax based on the total number of dwellings in the community. Unlike the French who ruled its colonies directly through assimilation, the British form of colonization was considered _____, and local leaders were cultivated and maintained tradition authority as a middle person between the British and the indigenous populations.

16. David Livingstone and Henry Stanley were both members of the _____, which was an organization founded for the "advancement of geographical science." This organization was tied closely to the colonial project in Britain.

17. The eradication of poverty and hunger as well as the promotion of gender equality and the empowerment of women are all part of a broader policy called the _____.

Activities & Problems

1. Based on Figure 5.3, list two of the primary mineral resources of the following countries:

   (a) Mali _____, _____

   (b) Ethiopia _____, _____

   (c) Angola _____, _____

   (d) Niger _____, _____, _____

   (e) Central African Republic _____, _____

2. Based on Figure 5.19, answer the following questions about colonization in Africa.

   (a) In what year was French West Africa created? _____ French Equitorial Africa? _____ Anglo-Egyptian Condominium? _____ Belgium Congo? _____

   (b) What were the names of Germany's colonial holdings? _____, _____, _____, _____

   (c) What were the two Portuguese colonies named? _____, _____ The Portuguese could not connect their colonial holdings across the continent because the _____ were able to move north from their original holdings in southern Africa and claim the area between the two Portuguese holdings.

   (d) The Chokwe's existing "state boundary" was divided between which colonial powers? _____, _____ And the Fulani? _____, _____, _____

3. Based on Figure 5.28, offer a brief discussion of the relationship between education and contraceptive use in Nigeria in 1990 and Botswana in 1988.

_____

_____

_____

_____

_____

_____

4. Examining Figure 5.6b, describe the climatic patterns of the African continent.

_____

_____

_____

_____

_____

_____

5. Using Figure 5.12b, offer a description of the spatial distribution of places of acute and moderate to great risk of desertification.

_____

_____

_____

_____

_____

_____

6. Perhaps even more acute than the North Africa and Middle East region is the problem of population density in Sub-Saharan Africa. In this question, you are going to: (a) calculate projected population densities for the countries of this region; (b) compare those calculations to the 2006 population density; and (c) examine the total percentage of change in population density over the two time periods.

   (a) Turn to the "2006 Population Data Sheet" located at the back of this text. On a separate piece of paper make the following columns in this order (you may want to use a spreadsheet program, such as Excel, for this exercise): (a) country name, (b) area of country, (c) projected population for 2050, (d) population mid-2006, (e) population density 2006, (f) projected population density 2050, (g) total increase in population density.

   (b) To calculate the population density for 2006 (column e) and 2050 (column f), you must do the following: divide the population mid-2006 and the projected population for 2050 by the surface area.

Example: Tanzania (2006) = 37.9 million/364,900 square miles

                                        = 104.12 people per square mile

                        Tanzania (2050) = 72.7 million/364,900 square miles

                                        = 199.23 people per square mile

(c) Now you can compare 2006 and 2050 projected population density values and fill in column g. We do this by calculating the percentage of change over time. This is done by setting up an equation to compare the two values. In the equation, the 2006 population density is taken to be 100%. We then calculate to see how much more dense it will be in 2015.

Example: Tanzania = pop density 2006/100 = pop density 2050/x, solve for x.

                        Tanzania = 104.12/100 = 199.23/x

                        Tanzania = (199.23 × 100)/104.12 = x

                        Pop density percent change = 191.35%

                        Population density in 2050 will be 191.35% of the population density in 2006.

(d) Calculate the percentage of change for all the countries.

(e) Which countries are going to have the highest percentage of change in population density by 2050? Why do you think that is the case?

_____

_____

_____

_____

_____

_____

_____

_____

_____

Review Exam

1. Describe the landform geography of the Sub-Saharan region.

_____

_____

_____

_____

_____

2. Describe the changing geography of AIDS cases in Africa between 1982 and 1997.

_____

_____

_____

_____

_____

3. Why is Africa called the "cradle of mankind"? Explain.

_____

_____

_____

_____

_____

4. What are some of the impacts that the Cold War had on the region?

_____

_____

_____

_____

_____

5. In what way does the colonial legacy still affect the relationship between politics and peace in the region? Explain with examples.

_____

_____

_____

_____

_____

6. Unlike southern Africa, which is dominated by Bantu-speaking peoples, the language family for Madagascar is _____.

   (a) Semetic-Hamatic

   (b) Nilotic

   (c) Malay-Polynesia

   (d) Afrikaaner

7. A person living in a Mediterranean climate zone in sub-Saharan Africa would live in which country?

   (a) South Africa

   (b) Namibia

   (c) Botswana

   (d) Lesotho

8. Numerous organisms act as vectors for the transmission of diseases. Which of the following diseases is not passed to humans by mosquitoes, flies, or snails?

   (a) malaria

   (b) schistomiasis

   (c) trypanosomiasis

   (d) HIV

9. Which of the following is true about the colonial experience in Sub-Saharan Africa?

   (a) Some African groups, such as the Ashanti, actively resisted European rule.

   (b) All African groups were pleased to have the advances offered by European colonizers.

   (c) European colonization did little to disrupt agricultural practices in the region.

   (d) Europeans were less interested in exploiting the mineral resources of the region than the previous ruling African kingdoms.

10. Which of the following policies is reminiscent of Portuguese rule in Congo?

   (a) indirect colonial rule with the active participation of local populations in political decision-making

   (b) an active interest in promoting the assimilation of elites and the populace more generally

   (c) ruthless control of land and labor with severe punishment for those who did not participate in public work projects

   (d) an active investment in the social infrastructure of the local population

11. Which of the following countries is not part of the political region called "the Sahel"?

   (a) Nigeria

   (b) Niger

   (c) Mali

   (d) Mauritania

12. Which of the following is not true of the Sahel region?

   (a) Traditional livelihoods included nomadic pastoralism.

   (b) During the rainy season, the area blooms into a grassland.

   (c) The region is marked by periodic droughts that have increased in recent years.

   (d) Droughts are natural and have nothing to do with human land use practices.

13. Which of the following is most accurate?

    (a) The urban population of Sub-Saharan Africa in 2000 was between 10 and 20 %.

    (b) The urban population of Sub-Saharan Africa in 2000 was between 21 and 30 %.

    (c) The urban population of Sub-Saharan Africa in 2000 was between 31 and 40 %.

    (d) The urban population of Sub-Saharan Africa in 2000 was between 41 and 50 %.

14. If you were living in the region dominated by the Songhai Empire between C.E. 1400 and 1590, you would be living in which subregion in Sub-Saharan Africa?

    (a) West Africa

    (b) East Africa

    (c) southern Africa

    (d) Central Africa

15. Programs that provide credit and savings to self-employed poor people who cannot get commercial loans so that they can start businesses on their own are often called _____.

    (a) sustainable development programs

    (b) social capital programs

    (c) indigenous technology programs

    (d) microfinance programs

For each of the following statements, write T for "True" or F for "False" on the space provided.

16. Diamonds are important commodities in Botswana._____

17. In January a large proportion of the African continent is impacted by dry, northeast trade winds._____

18. Malaria is found throughout the coastal areas of South Africa._____

19. Most places in Africa experienced colonization for hundreds of years._____

20. Apartheid policy included the Population Registration Act, which required all people in South Africa to be registered as either Bantu, colored, or white._____

21. There is no relationship between women's education and contraceptive use in Sub-Saharan Africa. _____

22. Internal migration in Sub-Saharan Africa is a result of individuals seeking work and of refugees fleeing famine, floods, and violent conflict._____

23. Despite Kenya's independence from British rule, its capital city, Nairobi, is still marked by high levels of racial and class segregation._____

24. Hunger and famine in the Horn of Africa is the result of both drought and war._____

25. A person educated in Algeria in the 1950s would likely speak French._____

Key Map Items

Identify the following features on Map 5.1 (countries and cities) and Map 5.2 (landforms and waterways). On your political map, Map 5.1, label each city by placing a dot to mark the approximate location of that place. When mapping some of the landforms, you may need to create some sort of symbol system that portrays the landform, such as a ^^^^^^^^^^^ for a mountain range.

Countries	Major Cities	Landforms	Waterways
As listed in Chapter 5 of your textbook (Figure 5.1)	• Abidjan • Adis Ababa • Brazzaville • Cape Town • Dakar • Dar es Salaam • Freetown • Harare • Johannesburg • Kinshasa • Lagos • Lusaka • Mogadishu • Monrovia • Nairobi • Timbuktu • Walvis Bay	• Chad Basin • Congo Basin • Djouf Basin • Kalahari Basin • Sudan Basin • Kalahari Desert • Namib Desert • Sahara Desert • Ethiopian Highlands • Great Rift Valley • Cape Fold Mountains • Ruwenzori Mtns. • Tibesti Mountains • Fouta Djallon Highlands • Horn of Africa	• Blue Nile River • Congo River • Niger River • Okavango River • Orange River • Senegal River • White Nile River • Zambezi River • Lake Chad • Lake Kariba • Lake Nyasa • Lake Tanganyika • Lake Victoria • Mozambique Channel • Gulf of Guinea

Mapping Colonial Legacies and Political Violence

Goal: The impact of European colonialism on this region cannot be underestimated. As the textbook highlights, one of the legacies of colonialism has been political violence. In this exercise, you are going to map out the pattern of European colonization at one point in time and layer onto that map a projection of Cold War Conflict and UN Peacekeeping missions. You will need to read the textbook, particularly pages 230–241 and Figures 5.19, 5.20, 5.23, 5.25, and 5.36.

1. Begin by sketching out in different colors the rough colonial boundaries in 1914 based on the modern-day political map (these boundaries are not always perfectly aligned but they are fairly close). Describe what pattern you see. Which countries are independent in 1914?

_____

_____

_____

_____

_____

2. Place on the map a square for each "conflict and intervention" in the region since the Cold War began. Make sure to place the square within the boundaries of the appropriate country.

3. Place a circle for each "UN peacekeeping mission" into the region. Make sure to place the circle within the boundaries of the appropriate country.

4. Describe the pattern that you see. Is there any correlation between European colonial powers and modern-day conflict? Explain.

_____

_____

_____

_____

_____

_____

_____

_____

_____

_____

_____

_____

5. What impact did colonialism have on the modern-day political geography of Sub-Saharan Africa that may have facilitated the conflicts that you have mapped? Explain.

_____

_____

_____

_____

_____

_____

_____

_____

_____

_____

_____

6. What regional organizations have emerged in the postcolonial period that might mitigate the affects of conflict in the region? If you were to map these supranational organizations onto the colonial map do you think you would see parallels between the two patterns? Explain.

_____

_____

_____

_____

_____

_____

_____

_____

_____

_____

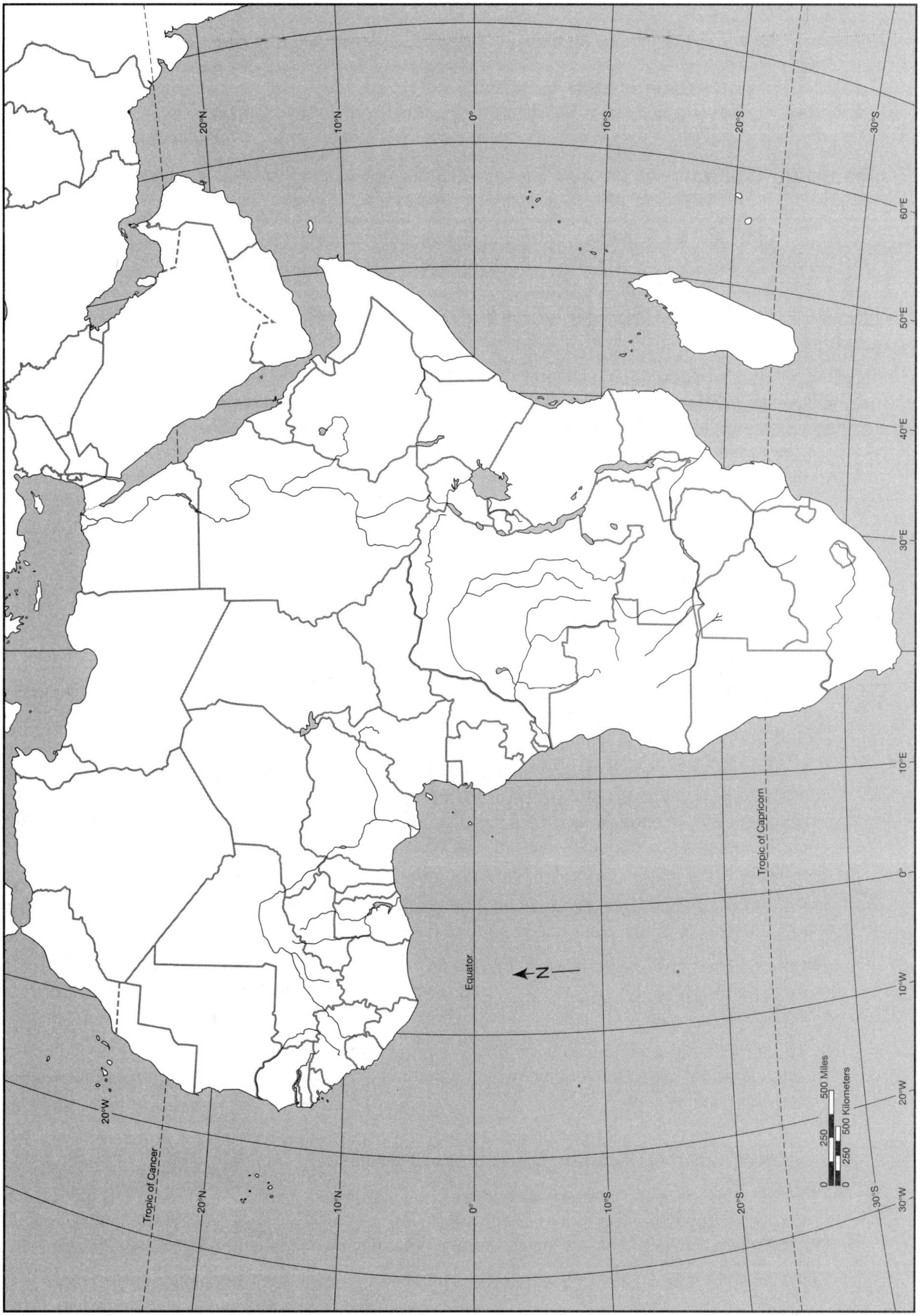

Map 5.3

CHAPTER 6

The United States and Canada

Learning Objectives

After reading and studying this chapter, you should be able to

1. understand the pre-Columbian historical geography of the region;
2. investigate the changing geography of the continent during the colonial and postcolonial periods, as well as the role of the United States as an imperial power;
3. have a working knowledge of the complex cultural geography of the North American region;
4. comprehend the position of North America in relation to the growing international interdependence of this region with other world regions;
5. examine urbanization and migration in their historical and spatial contexts;
6. compare and contrast the various subregional geographies of this region and the interplay between these subregions;
7. provide a description of the physical geographic processes that distinguish this region; and
8. consider the cultural, political, and economic contributions of this region to the processes of globalization.

Vocabulary Review

1. Between the Pacific coastal ranges and the Rocky Mountains lies a(n) _____ region, which consists of a set of basins, plateaus, and smaller ranges.

2. As Europeans moved into North America they brought with them cultural, political, and economic systems. This process, called _____, was a result of French, British, Spanish, and Danish influences during the early colonial period.

3. Even before the United States became independent, people born and raised in the region were becoming less loyal to colonial powers. Through _____, a process by which people in the Americas took on a distinct cultural and political identity, people in the North American colonies defined a new culture based on liberalism, individualism, capitalism, and Protestantism.

4. Prior to the introduction of a slave economy into North America, _____ from places such as Britain were used as labor in the colonies.

5. Canada has developed its natural resources very effectively over the centuries and thus remains an important _____, providing unprocessed natural resources for many other countries.

6. The forced and self-imposed segregation of ethnic and racial groups in North America has sometimes fostered misunderstanding, which can result in _____, such as anti-immigrant or racial violence.

7. _____ is a process by which peoples living within a particular territory (often a national border) develop cultural solidarity and similarity.

8. In Canada, unlike the United States, the official state policy is not one of assimilation. Instead, the Canadian government is more supportive of _____, whereby individual cultural groups are allowed to assert their cultural uniqueness in a local and national context.

9. While North America has grown based on immigration, the overall population of the region has also remained quite mobile. _____, which is the process of people moving within a national territory, is quite common for many people living in North America, as people move from one state or province to another.

10. North America, more so than almost any other region in the world, has witnessed a high level of _____ , as people continue to move to the fringes of urban areas.

11. In recent years, _____ has become the norm in many North American subregions as industrial employment moves outside the region seeking cheaper wages and less restrictive environmental policies.

12. Despite the decline in industrial labor, the changing capitalist system in North America has benefited from _____ , whereby older industrial regions are "dismantled" in order to develop new, innovative enterprises and regions.

13. In the 1980s, the agricultural economy began to change, and a _____ developed, leading to the financial failure and foreclosure of thousands of small family-owned farms.

14. Industrial growth in North America had economic benefits. Those benefits, however, have come at an environmental cost. _____ , which is the result of pollution generated from industrialization entering the atmosphere and returning to the ground through precipitation, has marked industrial regions, such as North America's rust belt.

15. In some areas, where industrial production developed largely unabated by legal sanctions, we now find _____ , sites with extreme levels of pollution that have yet to be remedied.

16. A _____ is a region of extreme urbanization that has developed within the early industrial corridor of North America around Boston–New York–Washington, D.C., in the United States. In Canada, a similar corridor developed, called _____ , between Windsor and Québec City.

17. _____ is a name for the process by which poorer urban neighborhoods are redesigned by a middle-class population, often forcing a rise in real estate prices and the further marginalization of poorer people in the immediate area.

18. In the extreme northern frontier of Canada, there exists a layer of _____ ,which means that it is impossible to sustain any agriculture.

19. Both Canada and the United States are _____ , which means that power is allocated to local political units.

20. In the _____ of the extreme northern frontier, where one can find a consistent layer of permafrost, there are still many animals and smaller forms of vegetation, such as mosses, lichens, and certain hardy grasses.

Activities & Problems

1. Using Figure 6.4 in your text and following the 40th parallel from the western coast of the United States to the eastern coast, trace the physiographic regions across the continent.

_____

_____

_____

_____

2. Examining Figures 2 and 3 on page 285 in your text, answer the following:

   (a) List the top four tobacco exporting countries in the world today. _____ ,
   _____ , _____ ,_____

   (b) List the top four tobacco-producing countries in the world today._____ ,
   _____ , _____ , _____

(c) Which regions are projected to have the highest increase in smokers between 1998 and 2008?

_____

(d) Which three countries will have the highest increase in the same time period?

_____ , _____ , _____ ,

_____ ,

3. Using Table 6.1, answer the following:

(a) Describe the pattern of change in labor force participation between 1961, 1991, and 2006.

_____

_____

_____

_____

_____

_____

(b) Explain why the pattern looks the way it does.

_____

_____

_____

_____

_____

_____

4. Using Figure 6.24 describe or explain the following:

(a) Describe the pattern of poverty.

_____

_____

_____

_____

_____

_____

(b) Offer some explanation as to why the map looks the way it does.

_____

_____

_____

_____

_____

_____

5. Using Figure 6.14, answer the following:

(a) Describe the waves of immigration into the U.S. between 1800–2000.

_____

_____

_____

_____

_____

_____

(b) Currently, which region provides the largest number of migrants? Why?

_____

_____

_____

_____

_____

_____

(c) What factors caused the flow of migrants into the U.S. to slow down between 1930 and 1940?

_____

_____

_____

_____

_____

_____

Review Exam

1. Offer a brief explanation as to how North America emerged as part of the "core" of the world economy by the end of the 19th century.

_____

_____

_____

_____

_____

_____

2. Describe the pattern of subsistence practices of Native Americans in North America prior to European colonization.

_____

_____

_____

_____

_____

_____

3. Despite a policy of "isolation" in the first half of the 20th century, the United States was actively aggressive beyond its boundaries. What sparked U.S. intervention in Latin America and the Caribbean?

_____

_____

_____

_____

_____

4. What social, political, and economic changes have prompted some in Québec to push for separation from the rest of Canada?

_____

_____

_____

_____

_____

_____

5. Briefly discuss the role of cities in the new economy. Examine how the development of new econo-my operated differentially across the subregions of the United States and Canada.

_____

_____

_____

_____

_____

_____

6. Which of the following is best supported by our current data/evidence on the earliest migration of humans to the North American continent?

   (a) The earliest any human appeared in North America was 11,000 years ago.

   (b) The earliest any human appeared in North America was 12,000 years ago.

   (c) The earliest any human appeared in North America was 15,000 years ago.

   (d) The earliest any human appeared in North America was 25,000 years ago.

7. Prior to 1492 it was possible that as many as _____ number of languages were spoken in North America.

   (a) 10   (b) 100   (c) 1000   (d) 2000

8. Immediately following colonization of North America by Europeans, the primary source of labor in the colonies was _____.

   (a) independent, free migrants from eastern Europe

   (b) slave labor from Africa

   (c) indentured servants from Europe

   (d) large groups of enslaved Native Americans

9. At the end of the 1800s the majority of migrants were coming from which world regions?

   (a) northern/western Europe and southern/eastern Europe

   (b) northern/western Europe and Asia

   (c) southern/eastern Europe and Asia

   (d) Latin America and Asia

10. Which of the following is true of the differences between migration patterns to the United States and Canada during the colonial period?

   (a) Most people migrating to Canada were from eastern Europe.

   (b) Most people migrating to Canada were British and French.

   (c) Most people migrating to the United States were British and French.

   (d) Most people migrating to the United States were German.

11. Which of the following has NOT had a significant impact on the process of suburbanization in North America?

(a) horse-drawn streetcars

(b) rail services

(c) automobiles

(d) airplanes

12. Which of the following is representative of policy put in place by the Canadian government to ward off the invasion of American culture?

(a) limiting the number of Americans that can cross into Canada each year

(b) levying a tax on *Sports Illu strated* because it did not contain enough material on Canadian sports

(c) building a large satellite system that can stop the flow of U.S. television into Canada

(d) mandating that 100% of all music on Canadian radio must be Canadian

13. The set of basins, plateaus, and small ranges lying between the Pacific Coastal Range and the Rocky Mountains is called the _____.

(a) high plateau range

(b) intermontane

(c) high fault zone

(d) great plains

14. The Pacific Rim refers to which of the following?

(a) all countries in the Americas that border the Pacific

(b) all countries that border the Pacific

(c) all countries in Asia that border the Pacific

(d) the United States, Japan, and China

15. Which of the following major Canadian cities is located closest to the U.S. border?

(a) Ottawa

(b) Montreal

(c) Québec

(d) Vancouver

For each of the following statements, write T for "True" or F for "False" on the space provided.

16. The United States and Canada are both members of the international Asia-Pacific Economic Cooperation organization._____

17. The North American core in the United States is called Main Street._____

18. Most people living in the northern frontier are active agriculturalists, planting crops throughout the year._____

19. Deindustrialization is often signaled be a rapid decline in manufacturing._____

20. About one-third of the population of Canada is Native American._____

21. Government policy in Canada has been supportive of monoculturalism, based on the philosophy that everyone should subscribe to one set of cultural values that is Canadian._____

22. The first major wave of internal migration in the region was marked by the movement of people of European descent from the eastern seaboard into the interior parts of the continent._____

23. The new regional economy of North America is marked by a recent and steep rise in industrialization._____

_____

_____

_____

_____

_____

_____

5. Do you see any regional clustering of poverty? Are there any states that are outliers relative to their neighboring states? Is there any consistent spatial pattern within the cluster of states with the highest rates? Why or why not?

_____

_____

_____

_____

_____

_____

_____

_____

6. Looking at the map of "Research and Development (R&D) expenditures and R&D/gross state product ratios by state 2001" on page 305 in your textbook, do you see a similar pattern in your map of poverty rates in the U.S.? Do you think there is a connection between the two patterns? If so, how so? If not, why not?

_____

_____

_____

_____

_____

_____

_____

_____

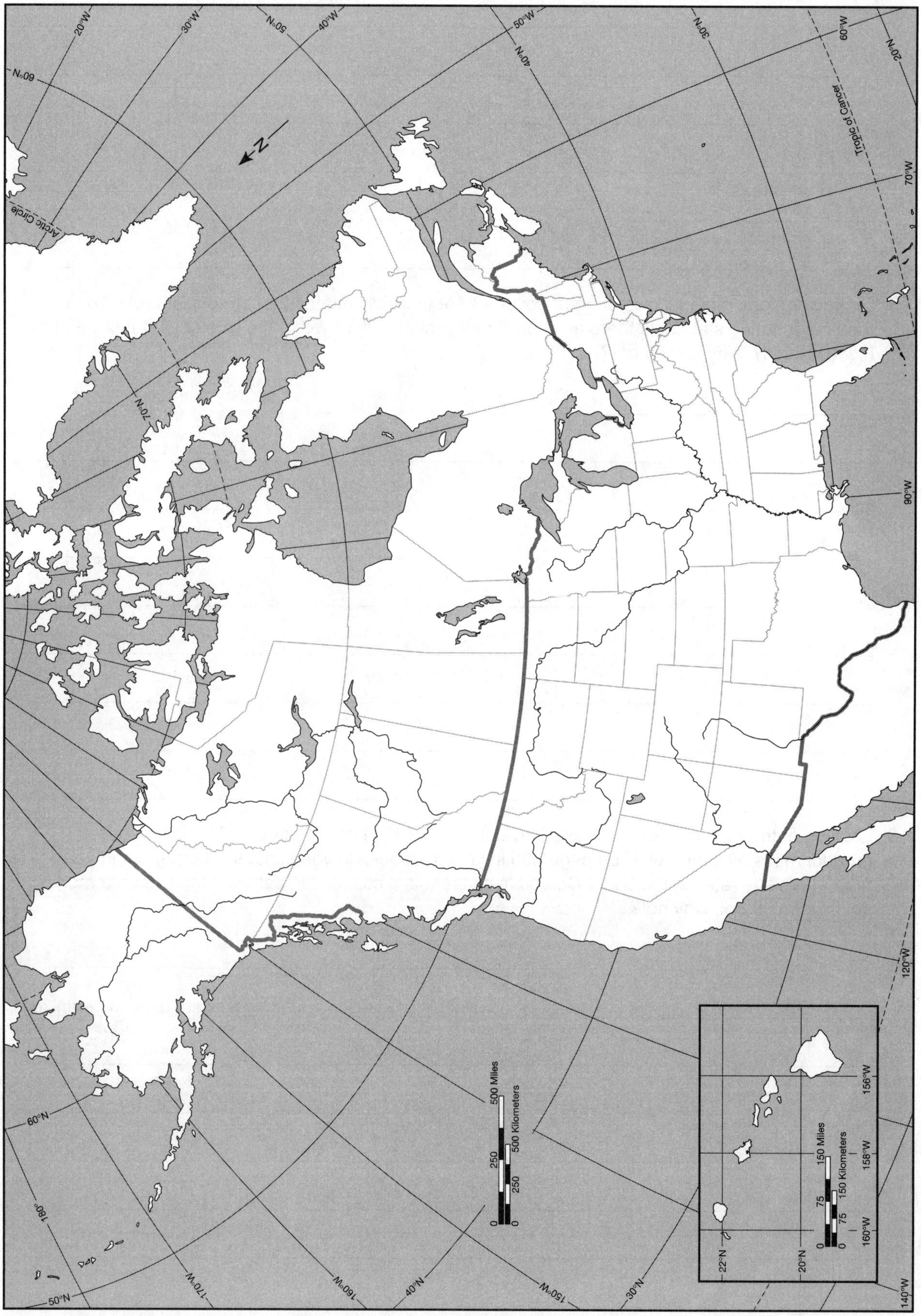

Map 6.3

CHAPTER 7

Latin America

Learning Objectives

After reading and studying this chapter, you should be able to

1. have a working knowledge of the unique climatic and physiographic subregions—including features such as altitudinal zonation and El Niño—of Latin America and the Caribbean;
2. understand the processes under which the peoples of the region transformed the physical environment;
3. comprehend the complex colonial and imperial history of the region beginning in 1492 and continuing up to the present day;
4. compare and contrast the multiethnic fabric and the syncretism that marks this region's complicated cultural geography;
5. examine political, economic, and social changes as they have impacted the region's rural and urban geographies;
6. have a working knowledge of the intricate relationship between humans and the environment in regional context;
7. comprehend the various push and pull factors that create the various migration patterns in and between areas in this region; and
8. understand the different primary urban areas in the region and their importance as political, economic, and social centers.

Vocabulary Review

1. _____ can be used to classify Latin America's mountain regions into four distinct climatic, vegetative, and human activity subregions.

2. The _____ is a theory that the Americas were untouched prior to European contact in 1492 and that native peoples lived in harmony with their environment.

3. In 1494, Pope Alexander VI drew a line of demarcation under the _____, at which time the world was divided between Spanish and Portuguese interests.

4. The term _____ denotes the process by which native populations in the Americas were decimated by diseases brought by Europeans.

5. Colonial powers in Latin America and the Caribbean established _____, large estates through which they raised cattle, wheat, olives, and fruit for sale in Europe.

6. _____ took the place of indigenous agricultural systems and were used to organize the production of single crops, such as sugar.

7. In the postcolonial period, local rulers took the place of colonial powers and maintained the existing systems of production, creating what some have called _____, countries that are run in the interest of multinational fruit corporations.

8. In the 1980s and 1990s, the economies of Latin America and the Caribbean became quite sluggish and some witnessed stagflation (a combination of economic stagnation and hyperinflation). In the wake of these economic crises, the International Monetary Fund imposed _____ under which governments were to reduce spending on subsidies and trade barriers as well as privatize government-owned industries in exchange for large-scale loans.

9. _____ is a term that was used to describe someone of Spanish/African descent during the colonial period, while _____ was used to describe someone of African/Indian descent.

10. In countries where one city dominates political, economic, and social life, we often call that _____.

11. _____ were given work permits to enter the United States from Mexico between 1942 and 1964. Many of these migrants never returned to Mexico.

12. A new form of Catholic practice, known as _____, focuses attention on the poor and disadvantaged segments of society. Practitioners mixed the ideas of Jesus Christ with the philosophies of Karl Marx.

13. One of the ways in which governments in the region tried to deal with inequity was to establish _____ programs under which land would be redistributed in attempts to increase productivity and decrease social unrest.

14. Many governments have moved away from growing and selling grains on the international market, instead focusing on _____, such as fruits, vegetables, and flowers.

15. People who do not work in waged jobs that are often taxed, might instead participate in the _____, which includes a variety of self-employed, income-generating activities.

16. The_____ was defined by historian Alfred Crosby to explain the interchange of crops, animals, and diseases between Americas and Europe and Africa.

17. A fairly recent form of economic development called_____ is intended to both provide local people with jobs while simultaneously protecting endangered ecosystems.

18. The hunt for new and unique tropical flora, known as _____, is driven by the search for new medicines and other commercial uses of rare plant life.

19. In 1823, the United States established the_____, which stated that any further European colonization or interference in the Western Hemisphere would be seen as an act of aggression.

20. Banks in the Bahamas and the Cayman islands offer_____, including tax-exempt and confidential banking.

Activities & Problems

1. Using Figure 7.7, trace the different climatic regions across South America at the 20th southern parallel.

   _____

   _____

2. Using Figure 7.8, list three primary commodities that one could find in the following altitudinal zones:

   (a) Tierra Caliente:_____ _____ _____

   (b) Tierra Templada:_____ _____ _____

   (c) Tierra Fria:_____ _____ _____

   (d) Tierra Helada: _____ _____ _____

3. Based on Figure 7.18, name the year in which the U.S. intervened in the following countries.

   (a) El Salvador _____

   (b) Chile_____

   (c) Argentina _____

   (d) Grenada _____

   (e) Guatemala _____

4. Based on Figure 7.27, name the language that is most commonly spoken in the following places:

    (a) the east coast of Brazil _____

    (b) northern Mexico near the US border_____

    (c) Haiti _____

    (d) Bahamas _____

    (e) Suriname _____

    (f) Guatemala _____

5. Using Figure 7.11 and the accompanying text, describe the geographic extent of the Maya, Aztec, and Incan empires. Name some of the major accomplishments of these early world historical societies.

_____

_____

_____

_____

_____

_____

6. Using Figures 7.3 through 7.15 and the accompanying text, respond to the following statements and questions related to the physical and resource geography of Latin America and the Caribbean.

    (a) Describe the broad physical regions and landforms of the region.

_____

_____

_____

_____

_____

    (b) Compare and contrast the varied landscapes that are a result of the complex relationship between landform and climatic zones in the region.

_____

_____

_____

_____

_____

_____

_____

_____

_____

_____

(c) Discuss the relationship between landform geography and mineral resource industries in the region.

_____

_____

_____

_____

(d) Explain the impacts of El Niño and La Niña on the region.

_____

_____

_____

_____

(e) What is meant by the term biodiversity? Why is this region such a rich source of biodiversity?

_____

_____

_____

_____

Review Exam

1. What are some of the reasons given for the collapse of Mayan civilization between C.E. 500 and C.E. 1000?

_____

_____

_____

_____

2. How did the Incas adapt to their environmental conditions?

_____

_____

_____

_____

_____

3. Compare and contrast the stark differences between rich and poor in the region today.

_____

_____

_____

_____

_____

4. What is liberation theology? What sparked this movement and what resulted from its inception?

_____

_____

_____

_____

_____

5. Briefly examine the relationship between the geological history the region and its mineral wealth.

_____

_____

_____

_____

_____

6. If you were living on the coast of Peru in Lima, you would most likely experience what type of climatic region?

(a) tropical rainforest   (b) tropical or subtropical desert

(c) tropical savanna and grassland   (d) midlatitude grassland

7. If you were in Brasilia, the capital of Brazil, you would most likely experience what type of climatic region?

(a) tropical rainforest   (b) tropical or subtropical desert

(c) tropical savanna and grassland   (d) midlatitude grassland

8. Despite interest by Spanish and Portuguese colonial powers to exploit local labor, efforts to use local labor were frustrated by the rapid decline of indigenous populations. This process is known as _____.

   (a) demographic explosion

   (b) demographic exchange

   (c) demographic collapse

   (d) demographic exposure

9. Which of the following is true about the exchange between Europe and Latin America and the Caribbean?

   (a) The Europeans introduced rats, pigs, and potatoes to Latin America.

   (b) The Europeans introduced rats, pigs, and corn to Latin America.

   (c) The Europeans moved corn and manioc from the Americas to Africa.

   (d) The Europeans moved pigs and rats from the Americas to Africa.

10. Which of the following countries received the largest amount of slave labor from Africa?

    (a) Brazil   (b) Cuba   (c) United States   (d) Argentina

11. Which of the following are you likely to find in the Teirra Helada?

    (a) poultry   (b) Vicuna   (c) sugarcane   (d) EU

12. During the colonial period, if you lived in Latin America or the Caribbean but you were born in Spain, you would be called which of the following?

    (a) criollos   (b) mulatto   (c) peninsulares   (d) zambo

13. The Caribbean Development Initiative (CDI) is a _____.

    (a) free trade association

    (b) command economy association

    (c) nontraditional economy cooperative

    (d) an indigenous commodities production program

14. A speaker of Quechua, a language indigenous to South America, would most likely live in which region?

    (a) central South America region

    (b) Caribbean region

    (c) central Mexico region

    (d) Andean region

15. Which of the following is true about people who live in the Andes?

    (a) Most people grow crops for the international market.

    (b) Most people grow potatoes close to the base of the mountains.

    (c) Most people grow agricultural products for subsistence or local consumption.

    (d) Most people use horses to move goods up and down steep mountain paths.

For each of the following statements, write T for "True" or F for "False" on the space provided.

16. Caribbean culture is strongly influenced by African cultural practices. _____

17. Flowers and other nontraditional agricultural products are grown on large company landholdings throughout the region. _____

18. A farmer growing potatoes in the Andes would have to grow this crop in the Tierra Fria or Tierra Helada altitudinal zones. _____

19. The term Latin America was coined by the Portuguese in the 17th century. _____

20. Most of the mineral wealth in Latin America is found in the old crystalline rock where crustal folds bring older rock near the surface. _____

5. What other factors, beyond the physiographic, help explain the climatic map of this region? Explain. Try to represent these processes on your map.

_____

_____

_____

_____

_____

_____

_____

_____

_____

Map 7.3

CHAPTER 8

East Asia

Learning Objectives

After reading and studying this chapter, you should be able to

1. have a working knowledge of the complex physiographic differences represented across the region;
2. understand East Asia's important role in world historical and geographical perspective;
3. compare and contrast the different political and social geographies of the countries of this region;
4. comprehend the importance of population policy as it relates to intraregional and interregional patterns of migration and ethnicity;
5. appreciate the processes of urbanization and counter-urbanization in and across the region;
6. develop an appreciation of the region's ethnic and cultural diversity as well as the historical practices that distinguish this region;
7. explore the importance of different economic models of development and the implementation of these models across the region; and
8. understand the ways in which East Asian societies are marked by syncretic processes and cultural blending.

Vocabulary Review

1. In the post-1868 period in Japan, the government supported capitalist monopolies, called _____, through public funding.

2. The _____ is a loosely defined region of countries that border the Pacific Ocean.

3. The Japanese archipelago is a series of volcanic islands that make up part of the larger _____, which girdles the Pacific Ocean.

4. Following the Opium War of 1839–1842, China was forced to open a series of _____, which included Canton and Shanghai, as well as cede the island of Hong Kong to the British.

5. _____ are business organizations that were established in Japan as part of post-WWII economic reforms.

6. The _____ include the countries of Japan, South Korea, Taiwan, and Singapore, a number of newly industrialized countries that have witnessed rapid economic growth.

7. In the 1970s, there was a brief period of _____, when businesses and people decided to escape congestion and inflated land prices in metropolitan Japan.

8. The philosophy that one can analyze the physical attributes of places and improve the flow of cosmic energy is called _____ in Chinese, something that is practiced widely in China. This practice is also known as _____.

9. Despite high levels of economic development, Japan, like most advanced capitalist nations, still has peripheral regions. These peripheral areas are characterized as _____, and are defined by selective out-migration, restricted investment, and limited employment opportunities.

10. The clustering of economic activities due to cost advantages, also known as _____, can be found throughout Japan.

11. The South Korean government facilitated capitalist development by supporting _____, a group of industrial conglomerates.

12. The Tibetan Plateau is a _____, a mountainous block of earth that is surrounded by faults and folds, creating a displaced unit of uplifted rock.

13. The Chinese language is a difficult one to learn, but the government has tried to simplify the language by reducing the number of characters used as well as adopting a new system, called _____ , for spelling Chinese words and names using the Latin alphabet.

Activities & Problems

1. Using Figure 8.4, answer the following questions:

   (a) What kind of vegetation would you find in Ulan Batar? _____

   (b) What kind of vegetation would you find on the Xizang Plateau? _____

   (c) What kind of vegetation would you find on the Daba Range? _____

   (d) What kind of vegetation would you find surrounding Pyongyang? _____

   (e) What kind of vegetation would you find surrounding Beijing? _____

   (f) What kind of vegetation would you find in China in the region bordering Vietnam?
   _____

2. Using Figure 8.12 in your text, describe the following.

   (a) Describe the extent of the Japanese Empire just prior to 1931.

   _____

   _____

   _____

   _____

   _____

   _____

   (b) Describe the extent of the Japanese Empire after 1931 and before 1933.

   _____

   _____

   _____

   _____

   _____

   _____

   (c) Describe the expansion of the Japanese Empire in 1933.

   _____

   _____

   _____

   _____

_____

_____

3. Using Figure 8.15 in your text, describe the pattern of population density in East Asia in 1995. What impact might the spatial distribution of population have on resource allocation in the region?

_____

_____

_____

_____

_____

_____

4. Using Figure 8.19, answer the following questions:

   (a) Where would we find the largest Chinese immigrant community outside of China in the world today? _____

   (b) From which places have 19th century Chinese immigrant communities largely vanished today?

   (c) Which two countries have been the largest recipients of Chinese immigration in Europe? _____ and _____

5. Human-environment interactions is one of the main areas of study in geography. As a budding geographer, you need to consider how humans have adapted to the physical environment, modifying that environment to their needs. To do this, you also need to understand the different environmental contexts in which people live. With this in mind, answer the following:

   (a) Using Figures 8.3 and 8.4, describe the differing physiogeographic regions and vegetation/climatic subregions of East Asia.

   _____

   _____

   _____

   _____

   _____

   _____

   (b) Explain what primary economic activities you are likely to find in each of the regions you described in 5(a).

   _____

   _____

   _____

   _____

(c) Describe the relationship between primary economic activity and population density one might find in each of these areas across the region based on your answers in (a) and (b).

_____

_____

_____

_____

_____

(d) Using Figure 8.6 in your text, do the following: (a) describe what you see; and (b) explain the relationship between the physical and human geographies in this particular place as well as how the spatial organization of this place might impact local social organization.

_____

_____

_____

_____

_____

_____

_____

_____

_____

_____

(e) Using Figure 8.7 in your text, do the following: (a) describe what you see; and (b) explain the relationship between the physical and human geographies in this particular place as well as how the spatial organization of this place might impact local social organization.

_____

_____

_____

_____

_____

_____

Review Exam

1. Name two of the most significant contributions of the Qin dynasty to classical world China.

_____

_____

_____

_____

_____

_____

2. Describe the urban geography of the Japanese imperial state under the Tokugawa regime.

_____

_____

_____

_____

_____

_____

3. In what two areas do we find the highest level of ethnic tension in China? What is the main cause of these tensions?

_____

_____

_____

_____

_____

_____

4. What is the "Great Leap Forward" and what impact did it have on Chinese society?

_____

_____

_____

_____

_____

_____

5. What impact has the liberalization of the Chinese economy had on rural-to-urban migration in China?

_____

_____

_____

_____

_____

_____

6. The Shang dynasty was responsible for which of the following?

   (a) establishing the first Bronze Age state

   (b) developing civilization in southern China

   (c) building the first walled cities in China

   (d) subdividing China into several smaller empires

7. In which of the following regions would one most likely find nomads today?

   (a) Japan's Pacific corridor   (b) South China

   (c) Manchuria   (d) Mongolia

8. Which of the following is true of the Meiji Restoration?

   (a) Japan moved from feudalism to industrial capitalism and the government actively supported capitalist monopolies called zaibatsu.

   (b) Japan moved from feudalism to industrial capitalism and the government actively supported capitalist monopolies called chaebol.

   (c) Japan moved from feudalism to socialism and the government actively supported capitalist monopolies called zaibatsu.

   (d) Japan moved from feudalism to socialism and the government actively supported capitalist monopolies called chaebol.

9. Which of the following is not true of Japan's postwar economic miracle?

   (a) There were exceptionally high levels of personal savings in Japan.

   (b) Japan was able to acquire new technologies from the United States.

   (c) The Japanese government fostered local industry through a series of tax concessions.

   (d) The Japanese relied heavily on their vast body of natural resources, such as coal and oil, to foster the growing industrial economy.

10. Which of the following is true of Xinjiang, a territory that occupies one-sixth of China?

   (a) The majority of the population is Christian.

   (b) The majority of the population is Muslim.

   (c) The majority of the population is Buddhist.

   (d) The majority of the population is animist.

11. Which of the following is true of Taiwan?

    (a) The Nationalist government set up the Republic of China on the island in 1949 after being forced out of the mainland by the communists.

    (b) The Nationalist government had little interest in land reform policies after arriving in Taiwan.

    (c) Taiwan gained full diplomatic status in 1971 after Nixon visited the People's Republic of China.

    (d) Taiwan's economic growth rate has stagnated since the 1970s.

12. What is meant by the term *juche*?

    (a) This is a term used in South Korea and represents a mix of Stalinist socialism, self-reliant nationalism, and a cult of personality.

    (b) This is a term used in North Korea and represents a mix of Stalinist socialism, self-reliant nationalism, and a cult of personality.

    (c) This is a term used in North Korea and represents a mix of state capitalism, self-reliant nationalism, and a cult of personality.

    (d) This is a term used in South Korea and represents a mix of state capitalism, self-reliant nationalism, and a cult of personality.

13. In which area of East Asia are you most likely to find wet rice agriculture?

    (a) the Tibetan Plateau

    (b) the central mountains and plateaus region

    (c) the southern continental margin

    (d) the northern continental margin

14. Which of the following is true of Japan's imperial period?

    (a) They successfully conquered the region around Shanghai in southern China.

    (b) They successfully conquered South but not North Korea.

    (c) They successfully conquered North but not South Korea.

    (d) They successfully conquered both North and South Korea.

15. Which of the following best defines Confucianism?

    (a) This philosophy was based on worshiping numerous gods.

    (b) This philosophy was based on ethics and the principles of matriarchy.

    (c) This philosophy was based on ethics and the principles of good governance.

    (d) This philosophy was based on family and matrilineal descent.

For each of the following statements, write T for "True" or F for "False" on the space provided.

16. Taiwan gained its full international diplomatic status in 1971._____

17. The Chinese were successful in repelling the British during the Opium War of 1839–1842._____

18. In the early years of the People's Republic of China Mao Zedong wanted a large population to exploit China's large territory._____

19. Japan has witnessed a growing trend where women in their mid-to late 20s refuse to marry; instead they live at home and spend their disposable income on luxury items._____

20. Pastoralism and nomadism has been the basis of life on the North China Plain for almost 5,000 years._____

21. In the post-WWII period in Japan's Pacific Corridor, industry shifted from cotton, silk, and other textile production to iron and steel production. _____

22. Kim Jong Il, the son of Kim Il Sung, is known by his people as "Great Leader." _____

23. Taiwan is considered to be one of the Asian Tigers along with Japan, North Korea, and Singapore. _____

24. The Chiang Jiang River basin is one of the most productive grain regions in China. _____

25. In recent years, North Korea has witnessed an explosion of food production that has led to food surpluses for the last five years. _____

Key Map Items

Identify the following features on Map 8.1 (countries and cities) and Map 8.2 (landforms and waterways). On your political map, Map 8.1, label each city by placing a dot to mark the approximate location of that place. When mapping some of the landforms, you may need to create some sort of symbol system that portrays the landform, such as a ^^^^^^^^^^^ for a mountain range.

Countries	Major Cities	Landforms	Waterways
As listed in Chapter 8 of your textbook (Figure 8.1)	• Biejing • Guangzhou • Hong Kong • Kyoto • Osaka • Pyongyang • Seoul • Shanghai • Tokyo • Ulan Bator • Xi'an	• Altay Mountains • Daba Range • Himalaya Mountains • Khingan Range • Kunlun Mountains • Pamirs Mountains • Qinling Range • Sayan Mountains • Tien Shan Mtns. • Yablonovyy Range • Hainan Island • Mongolian Plateau • Tibet Plateau • Yunnan-Guizhou Plateau • Sichuan Depression • Turfan Depression • Tarim Basin • Zunghaer Basin	• Bo Hai Bay • East China Sea • South China Sea • Yellow Sea • Sea of Japan • Huang He River • Chiang Jiang River • Mekong River • Xi River

Mapping Urbanization

Goal: You are now going to build on the map you made in Mapping Exercise 1 to spatially locate the region's main urban centers. For this exercise you will need the WPDS located at the back of this book and a set of colored pencils.

1. Begin by turning to Figure 8.1 (p. 376–377). Using that map, locate all cities in the region with a population over 1 million. Distinguish between cities that are over 5 million and those that are between 1 and 5 million by using different symbols (in a similar way to how the textbook portrays these differences).

2. Now turn to page 399 in your text and look at Figure 8.15. Using that figure, roughly outline and lightly shade all parts of this region that have a population of over 100 people per square kilometer (use a unique color pencil and then make a map legend and label the legend with the color that corresponds to high population density). Where are these cities located in relation to the overall population density of the region? Explain.

_____

_____

_____

_____

_____

_____

_____

_____

_____

_____

3. Now turn to page 380 in your textbook and look at Figure 8.3. Using that figure as a guide draw the rough outline of the various physiographic regions on your map (using a new colored pencil for each region making sure to add those colors and their explanation to your map's legend). Is there also a relationship between the physiographic regional map and the location of large urban centers? Explain.

_____

_____

_____

_____

_____

_____

_____

_____

_____

_____

4. What has facilitated the growth of urban centers in this region historically? Explain.

_____

_____

_____

_____

_____

_____

_____

5. How do these urban centers service their outlying hinterlands (surrounding subregions)? Explain.

_____

_____

_____

_____

_____

_____

_____

_____

_____

6. What role has China's changing policy of economic reform (in the post-1973 period) played in facilitating the growth of urban centers and the changing map of urbanization in China? Explain.

_____

_____

_____

_____

_____

_____

_____

_____

_____

_____

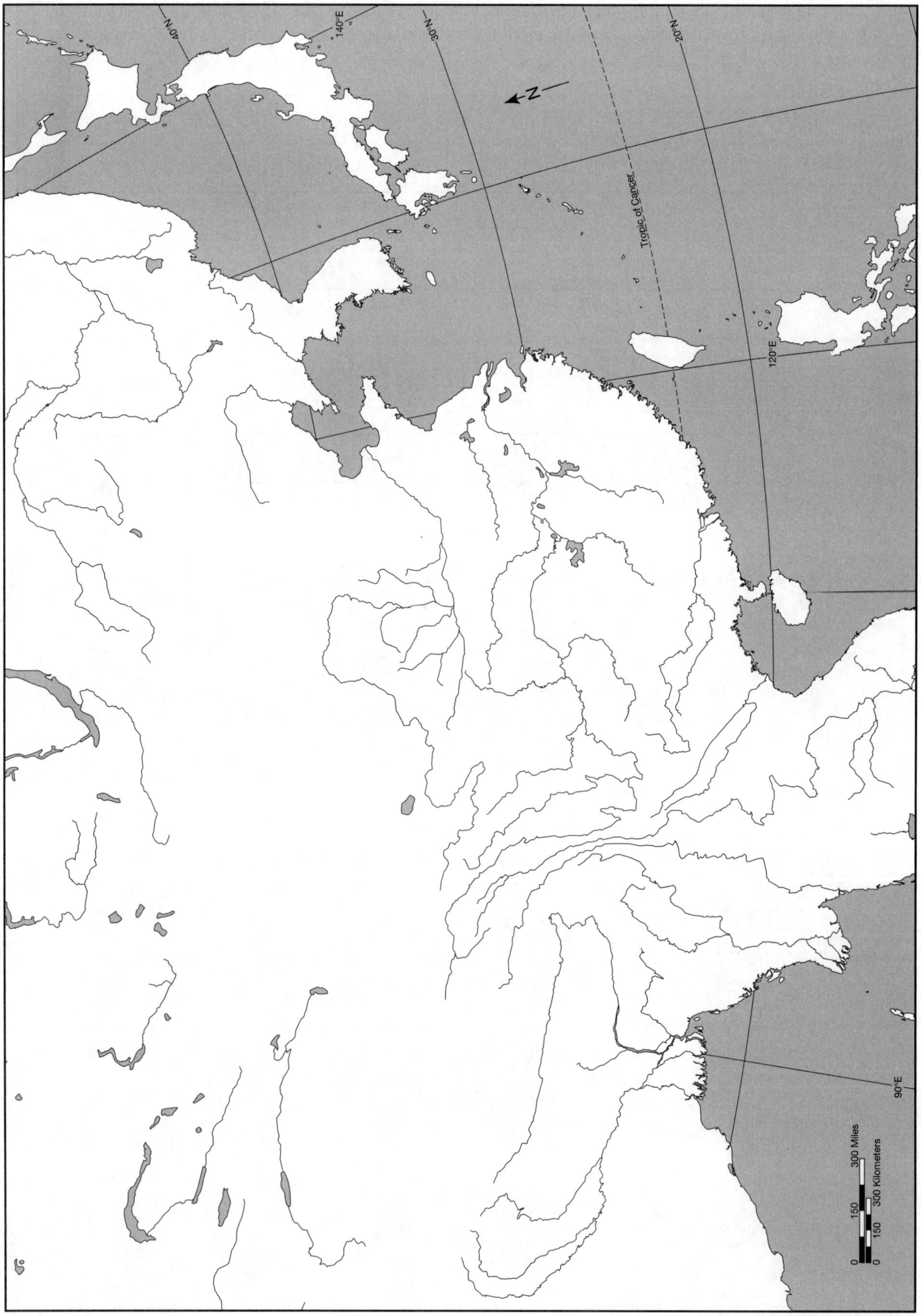

40°N
140°E
30°N
20°N
N
Tropic of Cancer
120°E
90°E
0 150 300 Miles
0 150 300 Kilometers

Map 8.3

CHAPTER 9

Southeast Asia

Learning Objectives

After reading and studying this chapter, you should be able to

1. have a working knowledge of the complex physiographic and climatic processes that mark this particular region;
2. understand the region's environmental history and how humans have adapted to this largely tropical set of environments;
3. compare and contrast the different colonial experiences of countries in the region as well as their current development trajectories;
4. comprehend the importance of the myriad ethno-cultural differences that mark the region, including the importance of overseas Chinese and other ethnic minorities;
5. appreciate the current political and economic tensions and the growing regional interdependence that has developed between historically juxtaposed countries;
6. develop an understanding of important regional and urban centers and their relationship with the current political and economic configuration of the region;
7. explore the relationship between religious practices and the gendered experiences of peoples in the region; and
8. understand the relationship between this region and the neighboring regions of South Asia and East Asia in historical perspective.

Vocabulary Review

1. All the countries of Southeast Asia are influenced by a _____ environment, with consistently warm temperatures and seasonal rainfall.

2. _____ defines the ecological division between Bali and Lombok, which kept several different species from mixing in the region.

3. Historically, wet rice, or _____, has been, and continues to be, the most important crop in Southeast Asia.

4. The islands of Indonesia, which were eventually colonized by the Dutch, were known to Europeans as the _____, because of their abundance of nutmeg, cinnamon, and cloves.

5. In 1830 the Dutch instituted the _____, whereby all Javanese people must contribute one-fifth of their land and labor to the production of export-oriented commodities.

6. While migration is often a "voluntary" process, in the 1950s the Indonesian government instituted the _____, which was designed to redistribute populations out of the densely populated island of Java and into the other islands of the Indonesian archipelago.

7. _____ populations were essential to the development of many of the colonial economies throughout Southeast Asia. Later, many of these migrants became part of the entrepreneurial class, establishing banks and insurance companies throughout Southeast Asia.

8. Singapore, a historical entrêpot, is the only city in Southeast Asia that can be considered a _____, similar to, perhaps, London and New York City.

9. Urbanization throughout Southeast Asia has been marked by _____, which means that population densities have outstripped employment opportunities.

10. In the last 50 years agriculture has been transformed by the twin processes of land reform and the _____, the latter of which focused on increasing yields of staple crops through the introduction of technological innovation.

11. A _____ is caused by an underwater earthquake and is defined by a series of large waves that move from deeper to shallow water sometimes making landfall.

12. Today, all 10 countries in Southeast Asia belong to the supranational organization called _____, which is most commonly referred to by its acronym _____.

13. The _____ refers to the United States' belief that if Vietnam were to be occupied by communist forces the rest of Southeast Asia would soon succumb to communism.

14. In the 1960s, most of the countries of Southeast Asia subscribed to the economic policy of _____ industrialization, seeking to develop domestic productive capabilities while limiting imports.

15. Unlike many other countries in Southeast Asia, Malaysia in the 1970s established _____, which is also sometimes called a _____, encouraging foreign capital to enter Malaysia to establish various industrial programs.

16. During the last ice age the Sunda Shelf was exposed, forming a _____ between Borneo and mainland Southeast Asia.

17. _____, which is also known as slash and burn agriculture, has historically been used to clear tropical rainforest areas for cultivation.

18. The Indonesian Philippine islands are part of the _____, which surrounds the Pacific and includes the volcanic islands of Japan as well.

Activities & Problems

1. Using Figure 9.7, answer the following questions.

   (a) How many dry months does Bangkok experience? _____

   (b) How many dry months does Kuala Lumpur experience? _____

   (c) How many dry months does Bandar Seri Begawan experience? _____

   (d) How many dry months does Dili experience? _____

   (e) How many dry months does Singapore experience? _____

2. Using Figure 9.8, describe the vegetation geography of Southeast Asia. What impact have humans had on the overall ecology of the region?

_____

_____

_____

_____

_____

_____

_____

_____

_____

_____

3. Using Figure 9.10, answer the following:

   (a) Which country has the most forest cover in total hectares? _____

   (b) Which country has the least forest cover in total hectares? _____

   (c) Which country has the highest deforestation rate between 1973 and 1985?
   _____

   (d) Which country has the lowest deforestation rate between 1973 and 1985?
   _____

   (e) Which country has the highest deforestation rate between 1990 and 2000?
   _____

4. Using Figure 9.12, name the countries that were colonized by the following European powers.

   (a) French _____

   (b) British _____

   (c) Dutch _____

   (d) Spanish _____

   (e) United States _____

   (f) According to the map, in what year were the Philippine Islands first colonized?
   _____ In what year did the colonial power in the Philippines change hands?
   _____

5. Using Figure 9.14, describe the pattern, scope, and extent of Japanese occupation of Southeast Asia during WWII.

   _____

   _____

   _____

   _____

   _____

   _____

Review Exam

1. Describe the different ways in which monsoons impact the mainland and insular regions of Southeast Asia.

   _____

   _____

   _____

   _____

2. Describe the relationship between population density and deforestation in Southeast Asia. What areas have witnessed the highest levels of deforestation in the region? Which regions have witnessed the lowest levels?

_____

_____

_____

_____

_____

_____

_____

_____

_____

_____

3. Describe Indonesia's transmigration program. What are some of the problems with the program?

_____

_____

_____

_____

_____

4. What role has overseas Chinese played in the political and economic development of Southeast Asia?

_____

_____

_____

_____

_____

5. What world regions have received the largest number of labor migrants from Southeast Asia? What kinds of jobs are these people doing? How is the work gendered?

_____

_____

_____

_____

_____

_____

6. Which of the following is not a conflict zone in Southeast Asia?

   (a) the Muslim rebellion in Mindanao, Philippines

   (b) the Shan rebellion in northern Burma

   (c) the Thai Muslim rebellion near the Laos border in northeastern Thailand

   (d) the Karen rebellion along the Thai/Burma border

7. Which three Southeast Asian countries constitute the "Golden Triangle"?

   (a) Burma, Thailand, and Laos   (b) Burma, Thailand, and Malaysia

   (c) Thailand, Laos, and Vietnam   (d) Thailand, Laos, and Cambodia

8. Which of the following statements best reflect the physical geography of Southeast Asia?

   (a) Burma and Thailand are part of the Ring of Fire.

   (b) The Philippines is in a major typhoon track.

   (c) Burma is subject to extreme earthquakes.

   (d) Most of Laos is at sea level.

9. Which of the following best describes the culture system?

   (a) This system required all Javanese to learn Dutch.

   (b) This system required all Javanese to wear Western dress.

   (c) This system required all Javanese to promote Dutch culture.

   (d) This system required all Javanese to devote one-fifth of their land and labor to export crop production.

10. Which of the following best describes what happened almost immediately in Vietnam after the re-unification of the country in 1975?

   (a) Over 2 million people fled North Vietnam for fear of persecution, and the United States supported economic development through aid projects.

   (b) Over 2 million people fled South Vietnam for fear of persecution, and the United States supported economic development through aid projects.

   (c) Over 2 million people fled South Vietnam for fear of persecution, and the United States established an embargo of Vietnam that lasted until 1993.

   (d) Over 2 million people fled North Vietnam for fear of persecution, and the United States established an embargo of Vietnam that lasted until 1993.

11. Overseas Chinese communities are best described by which of the following statements?

    (a) In most cases, Chinese have assimilated into all host societies with little problem.

    (b) Overseas Chinese faced high levels of discrimination at the hands of Japanese occupational forces in Southeast Asia during WWII.

    (c) Singapore and Malaysia became one country in 1982 to unite ethnic Chinese in Singapore with ethnic Malays in Malaysia under one common political and economic system.

    (d) Ethnic Chinese make up only a small percentage of the population of the island of Singapore today.

12. Which of the following is the most accurate statement about Southeast Asia's overall economic level?

    (a) Southeast Asia boasts some of the highest and some of the lowest gross national products in the world.

    (b) Overall the economies of Southeast Asia are all relatively equivalent.

    (c) The former socialist countries have enjoyed the most rapid economic growth since 1975 in the region.

    (d) Ethnic minorities in Southeast Asia, such as hill tribes in the mainland countries, are more likely to enjoy economic success than the majority populations.

13. Which of the following best represents the Mekong basin?

    (a) Hanoi is served by this large river.

    (b) The headwaters begin in China and end in the South China Sea.

    (c) There has been little dialogue between the countries in the basin region.

    (d) Since the river is slow there is little potential for hydroelectric power.

14. A person living in northern Vietnam would most likely practice which religion?

    (a) Hinduism   (b) Therevada Buddhism

    (c) Islam   (d) Mahayana Buddhism

15. Which of the following best reflects Spanish reasons for colonizing the Philippines?

    (a) They wished to spread Catholicism and gain access to spice trade routes.

    (b) They wished to spread Catholicism and stop the flow of spices out of the region.

    (c) They wished defend local religious practices and facilitate the growth of indigenous political systems in the region.

    (d) They wished to defend local religious practices and expand the Spanish Empire.

For each of the following statements, write T for "True" or F for "False" on the space provided.

16. The British colonized Thailand in the 19th century. _____

17. The concept deva-raja, or god-king, was brought to Southeast Asia from India. _____

18. Someone living in Java would most likely practice Islam. _____

19. Burmese is an Austro-Asiatic language. _____

20. Japan is often blamed for deforestation in Southeast Asia because of its high demand for hardwoods. _____

21. Bali is the only place in Southeast Asia where Hinduism is practiced. _____

22. In most cases, Southeast Asia is marked by a high number of extended families. _____

23. Only about 5% of the population of Southeast Asia currently lives in an urban area.
   _____

24. Thailand has been a net recipient of refugees in the last 20 years. _____

25. Under British control, Malaysia largely produced rubber and tin. _____

Key Map Items

Identify the following features on Map 9.1 (countries and cities) and Map 9.2 (landforms and waterways). On your political map, Map 9.1, label each city by placing a dot to mark the approximate location of that place. When mapping some of the landforms, you may need to create some sort of symbol system that portrays the landform, such as a ^^^^^^^^^^^^ for a mountain range.

Countries	Major Cities	Landforms	Waterways
As listed in Chapter 9 of your textbook (Figure 9.1)	• Bandar Seri Begawan • Bangkok • Cebu City • Chiang Mai • Hanoi • Ho Chi Minh City • Hue • Jakarta • Kuala Lumpur • Luang Prabang • Mandalay • Manila • Penang • Phnom Penh • Rangoon • Singapore • Vientiane	• Arakan Mountains • Annam Mountains • Barisan Mountains • Moluccas • Sulawesi • Borneo • Sumatra • Java • Bali • Timor • Mindanao • Isthmus of Kra	• Chao Phraya River • Irrawaddy River • Mekong River • Red River • Tonle Sap Lake • Straits of Malacca • Andaman Sea • Banda Sea • Celebes Sea • Java Sea • South China Sea • Sulu Sea • Indian Ocean • Gulf of Thailand

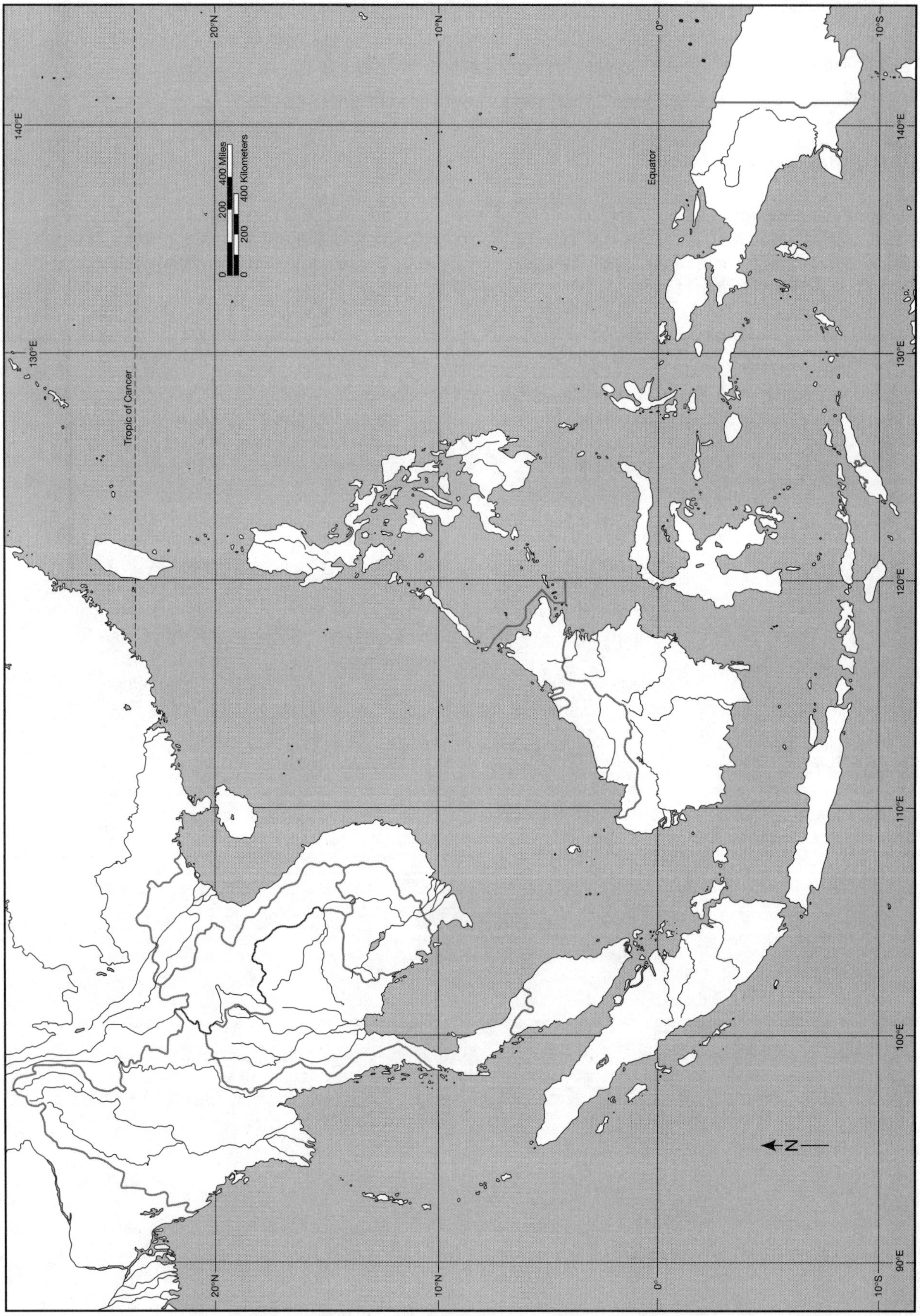
20°N
10°N
0°
10°S
140°E
130°E
120°E
110°E
100°E
90°E
Tropic of Cancer
Equator
0 200 400 Miles
0 200 400 Kilometers
N

Map 9.1

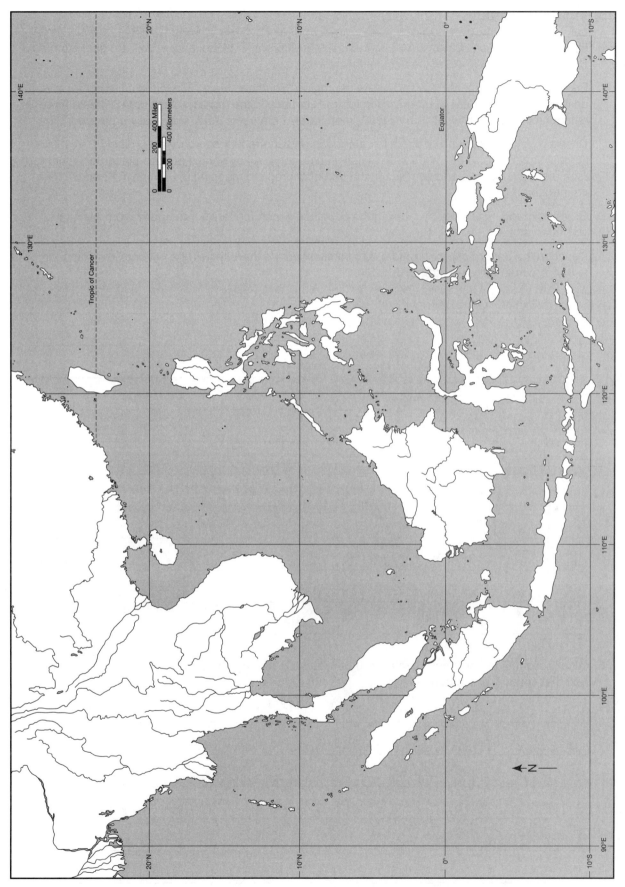

Map 9.2

Mapping Urbanization

Goal: Urbanization across Southeast Asia is dramatically different. In this exercise, you are going to map the "% Urban," using the WPDS at the back of this workbook.

1. Begin by turning to the WPDS and locate the region "Southeast Asia." Go to the column "% Urban," and copy down the rates of urbanization for each country in the region in rank order. Create three categories of data for "highly urbanized," "moderately urbanized," and "marginally urbanized."

2. Create a key for each one of the data categories and shade that category accordingly (i.e., the lightest shade should be used for the smallest category). A key should also be placed on your map (see discussion above and samples in the text of choropleth mapping in Figures 1.12 and 1.15 in your textbook).

3. Create a map of your data by labeling the countries according to your categorization. Also, give your map a title.

4. Describe the pattern of urbanization. What limitations are there to using a national level map for analyzing urbanization patterns? Explain.

_____

_____

_____

_____

_____

_____

_____

_____

_____

_____

_____

5. What accounts for the different rates of urbanization in the region? Explain with examples from the text that support your argument.

_____

_____

_____

_____

_____

_____

_____

_____

_____

_____

_____

6. Using Figure 9.1, identify all cities with a population of 1 million or more people. Place all those cities on your map. Is there a relationship between these large urban centers and the map of overall urban density? Explain.

_____

_____

_____

_____

_____

_____

_____

_____

_____

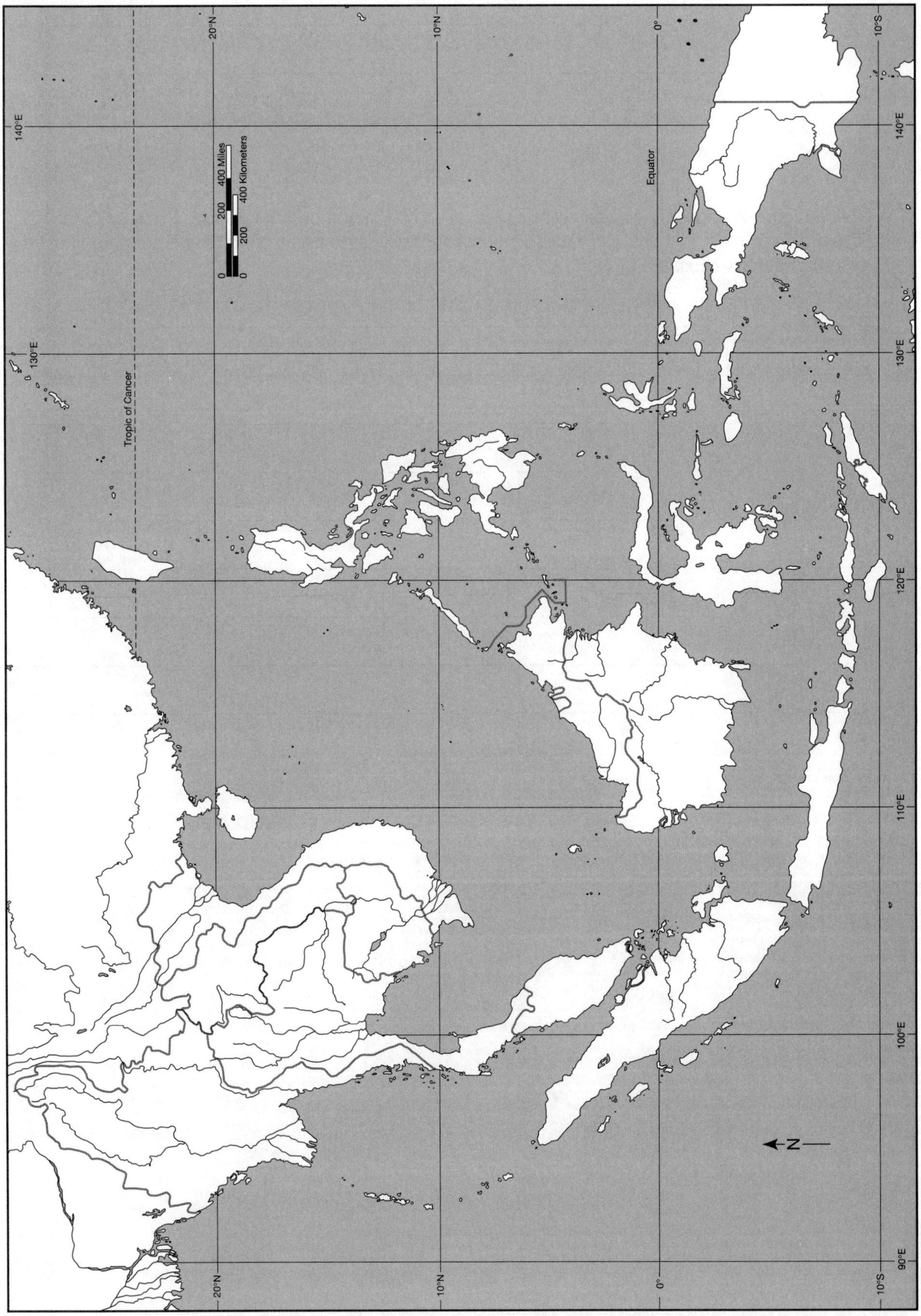
140°E
130°E
120°E
110°E
100°E
90°E
20°N
10°N
0°
10°S
Tropic of Cancer
Equator
0 200 400 Miles
0 200 400 Kilometers
N

Map 9.3

CHAPTER 10

South Asia

Learning Objectives

After reading this chapter, you should be able to

1. have a working knowledge of the complex physiographic differences that are represented across the region;
2. understand South Asia's historical geography and the impact of that geography on the everyday lives of people in the region;
3. compare and contrast the different social, economic, and political development paths of key countries in the region;
4. comprehend the importance of language and religion in marking the complex political geography of this region;
5. appreciate the importance of poverty and its impact on women and children in the region as well as the goals of alternative development strategies in the region;
6. develop an appreciation of the region's ethnic and cultural diversity;
7. explore the importance of the various subregional geographies that distinguish South Asia; and
8. understand the ways in which South Asian societies cope with the changing environmental conditions and hazards in the region.

Vocabulary Review

1. In 1876, after Queen Victoria was declared Empress of India, the British established the _____, which by 1890 extended their control over the entire region, excepting present-day Afghanistan and Nepal.

2. The Himalayan and Karakoram Mountain ranges are main features of the physiographic region known as the _____.

3. The Taj Mahal is an important symbolic feature of the landscape in India today. It is a legacy of the _____ Empire, which promoted the spread of Islam in South Asia beginning in the 15th century.

4. Throughout most of India, people practice _____, a polytheistic religion that finds its roots in early Dravidian and Indo-Aryan societies.

5. The center of India's massive movie market is the area known as _____ in Bombay.

6. _____ is a system of kinship groupings, or jati, that are reinforced by language, region, and occupation.

7. Harappan agriculturalists established large-scale civilization in the _____, an area where they developed irrigation systems almost 5000 years ago.

8. The South Asian _____ in the 19th century was marked by a significant movement of South Asians to various British colonies in East Africa, South Africa, and Southeast Asia.

9. South Asia's mountain ranges produce a strong _____, causing moist air from the sea to lift and condense. This process produces heavy rainfall in different parts of the region in both the summer and winter monsoon seasons.

10. Situated between the Western and Eastern Ghats, the _____ is home to India's technology corridor centered around the city of Bangalore.

Activities & Problems

1. Using Figure 10.16, answer the following:

   (a) Name the 10 dependent states of South Asia under British rule.

   _____

   _____

   _____

   _____

   _____

   _____

   (b) Describe the pattern of annexation between 1753 and 1805 and again after 1805. Consider where that annexation took place.

   _____

   _____

   _____

   _____

   _____

   _____

   (c) Based on the map, which two present-day countries remained free from British rule?
   _____ and _____

2. Using Figure 10.20, describe the pattern of major religions in South Asia.

   _____

   _____

   _____

   _____

   _____

3. Using Table 10.1, answer the following questions.

   (a) In which country is it most likely that a person would not have access to safe water?

   _____

   (b) In which country does the adult illiteracy rate exceed 60%? _____

   (c) In which countries does the percentage of the population without access to sanitation exceed 65%? _____ , _____ , and _____

4. Using Figure 10.3, answer the followig:

(a) Describe the physiographic subregional geography of South Asia.

_____

_____

_____

_____

_____

_____

_____

(b) In which of the following four physiographic subregions would you find:

(i)   Thar Desert _____

(ii) Deccan Plateau _____

(iii) Khasi Hills _____

(iv) Western Ghats _____

(v) Balochistan _____

5. (a) What prompted the Soviet Union to invade Afghanistan in 1979?

_____

_____

_____

_____

_____

_____

(b) In the post-Soviet period, what political problems have made maintaining peace in Afghanistan so difficult?

_____

_____

_____

_____

_____

_____

_____

_____

Review Exam

1. Compare and contrast the physical features of the plains and the coastal fringe.

   _____

   _____

   _____

   _____

   _____

   _____

2. Describe the monsoon patterns and the impact of the monsoons on the agricultural geography of South Asia.

   _____

   _____

   _____

   _____

   _____

   _____

3. What are the main causes of ethnic tension in Sri Lanka? Explain.

   _____

   _____

   _____

   _____

   _____

   _____

4. Describe the uneven geography of capitalist development in India.

   _____

   _____

   _____

   _____

   _____

5. Describe the in-migration of the Aryan peoples into South Asia. Where did they settle and what impact did they have on the landscape?

_____

_____

_____

_____

_____

_____

6. Which of the following two countries are nuclear powers?

    (a) India and Afghanistan   (b) India and Sri Lanka

    (c) Pakistan and Afghanistan   (d) Pakistan and India

7. The Western Ghats is marked by which of the following?

    (a) relatively high population densities and predictable yearly rainfall

    (b) relatively high population densities and unpredictable yearly rainfall

    (c) relatively low population densities and predictable yearly rainfall

    (d) relatively low population densities and unpredictable rainfall

8. Which of the following is true of Harrappan society?

    (a) It emerged in the Indus River Valley 4500 years ago and eventually opened up trading with Mesopotamia.

    (b) It emerged in the Indus River Valley 4500 years ago and survived even after floods covered its major cities in mud and silt.

    (c) It emerged in the Ganges River Valley 4500 years ago and eventually opened up trading with Mesopotamia.

    (d) It emerged in the Ganges River Valley 4500 years ago and eventually moved the entire civilization to southern India after environmental disasters made it impossible to live in the north.

9. Under the Mughal Empire and Akbar's rule, in India one could find all of the following except _____.

    (a) Persian as the official language

    (b) princely kingdoms kept intact

    (c) taxing on non-Muslims

    (d) an increased status of Islam in the region

10. Pakistan, which means "land of the pure," was established as a result of what geopolitical factor?

    (a) the partition of South Asia by the British based on ethnic and religious groupings

    (b) the partition of South Asia by Gandhi based on ethnic and religious groupings

    (c) the partition of South Asia into Islamic and Buddhist regions

    (d) the partition of South Asia into Hindu and Buddhist regions

11. Which of the following languages would a Buddhist most likely speak in Sri Lanka?

    (a) Hindi   (b) Tamil   (c) Sinhalese   (d) Bengali

12. Which of the following is most true about the economy in South Asia?

   (a) Women are more likely to have access to education than men.

   (b) Women are more likely to participate in the informal economy than men.

   (c) Women are more likely to participate in the formal economy than men.

   (d) Women have fewer restrictions on public action than men.

13. When someone has no collateral to back up a loan they might have to enter into what type of economic relationship?

   (a) indentured servitude   (b) bonded labor

   (c) free labor   (d) slave labor

14. Which of the following is most true of Sri Lanka?

   (a) It was granted independence by the British in 1997 at the same time that they handed Hong Kong over to China.

   (b) It continues to be plagued by ethnic and religious tensions between Tamils and Sinhalese.

   (c) It is home to a large Islamic population.

   (d) Almost 80% of the population does not have access to safe water.

15. Which of the following is true of Mumbai?

   (a) It is one of the smallest cities in South Asia.

   (b) It is considered to be an economic drain on India.

   (c) It is a center for agricultural production.

   (d) It is the historical headquarters of the British East India Company.

For each of the following statements, write T for "True" or F for "False" on the space provided.

16. It is possible that families in need of cash might sell their child into bonded labor in South Asia._____

17. Only 25% of the region's population lives in rural villages today._____

18. In recent years, India has developed a large, thriving middle class._____

19. Poverty can be found throughout both rural and urban South Asia._____

20. The Portuguese successfully established a colony in the region of South Asia known as Goa in the 1600s._____

21. Bangladesh was previously known as East Pakistan._____

22. Informal economic activities, such as street vending, are rarely seen on the streets of urban South Asia._____

23. In rural South Asia schooling is prolific and literacy rates are quite high._____

24. Under the Raj system, British rule was extended and plantation agriculture was expanded throughout South Asia._____

25. Almost 80% of South Asia remains forested today._____

Key Map Items

Identify the following features on Map 10.1 (countries and cities) and Map 10.2 (landforms and waterways). On your political map, Map 10.1, label each city by placing a dot to mark the approximate location of that place. When mapping some of the landforms, you may need to create some sort of symbol system that portrays the landform, such as a ^^^^^^^^^^^^^ for a mountain range.

Countries	Major Cities	Landforms	Waterways
As listed in Chapter 10 of your textbook (Figure 10.1)	• Bangalore • Chennai • Colombo • Delhi • Dhaka • Hyderabad • Islamabad • Kabul • Karachi • Kathmandu • Lahore • Mumbai • Srinigar • Thimphu	• Aravalli Hills • Himalaya Mtns. • Hindu Kush Mtns. • Sulaiman Range • Karakoram Range Kirthar Range • Deccan Plateau • Thar Desert • Eastern Ghats • Western Ghats	• Brahmaputra River • Ganga River • Godavari River • Indus River • Krishna River • Narmada River • Yamuna River • Arabian Sea • Indian Ocean • Bay of Bengal

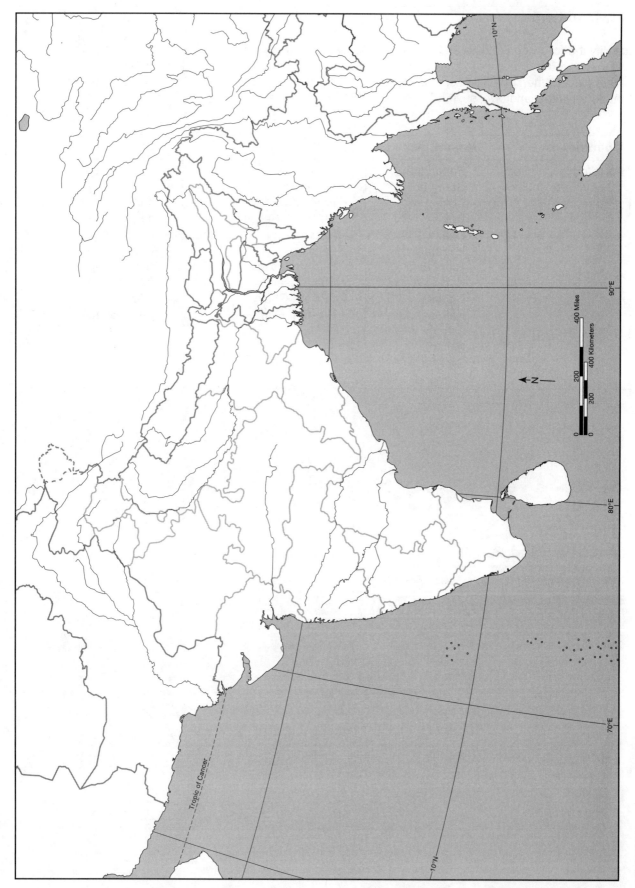

Map 10.1

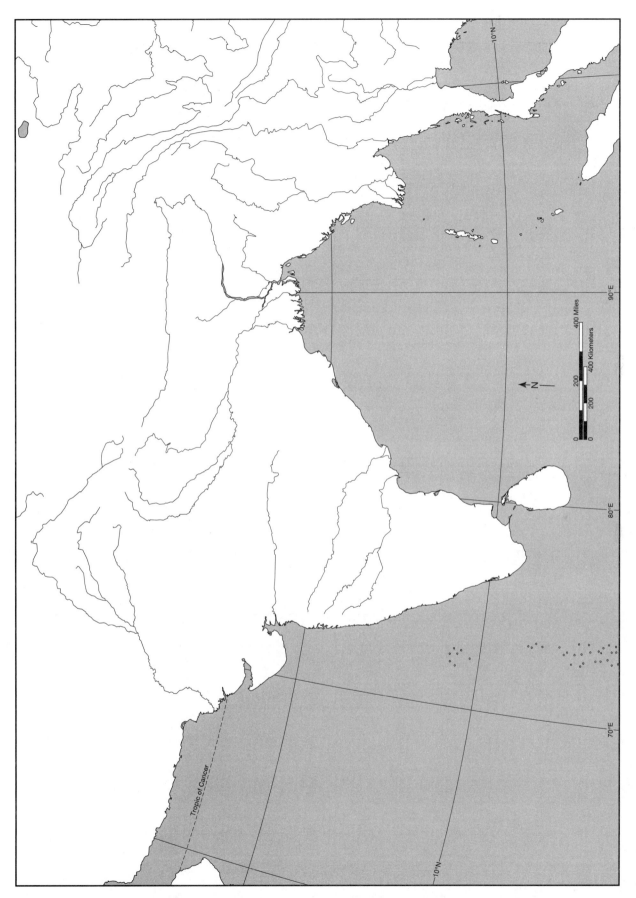

Map 10.2

Geopolitical Conflict

Goal: In this exercise, you are going to map the significant geopolitical conflict areas in the region and analyze the spatial distribution of those conflicts.

1. Read through Chapter 10 and highlight discussions regarding current geopolitical and or ethnic conflict in the region. Using your imagination, come up with an icon that best represents ethnic conflict in the region. On the blank political map place your icons in proximity to the conflict. You may need to create a boundary and/or an explanatory box to highlight the details of the conflict (see example in Chapter 5 on Sub-Saharan Africa, Figure 5.25).

2. What pattern emerges? Are these conflicts concentrated in any particular part of the region?

_____

_____

_____

_____

_____

_____

_____

3. What geopolitical and/or cultural and ethnic factors promote conflict in the region? (e.g., think about borders and cultural distributions). Explain.

_____

_____

_____

_____

_____

_____

_____

_____

_____

_____

_____

_____

4. What impact did European colonialism have on the region? What effect does that colonial legacy have on the map of modern-day conflict you have created? Explain.

_____

_____

_____

_____

_____

_____

_____

_____

_____

_____

_____

_____

_____

5. In what way is the current map of conflict complicated by the modern-day development of nuclear weapons in the region? Explain.

_____

_____

_____

_____

_____

_____

_____

_____

_____

_____

_____

6. What role do international forces, such as the U.S., play in the current regional conflicts? Explain.

_____

_____

_____

_____

_____

_____

_____

_____

_____

_____

_____

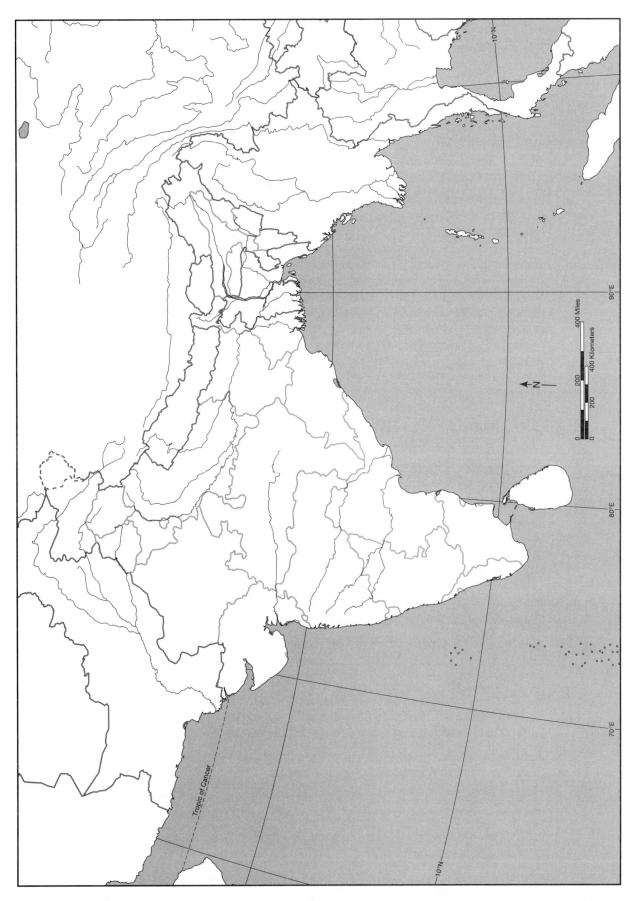

Map 10.3

CHAPTER 11

Australia, New Zealand, and the South Pacific

Learning Objectives

After reading and studying this chapter, you should be able to

1. have a working knowledge of the climatic and landform geographies of this expansive world region;
2. understand the complex relationship between indigenous populations and recently transplanted European peoples;
3. have a working knowledge of the ethnic and political geography of this largely insular region;
4. understand the environmental politics that have marked this region in the 20th and 21st centuries;
5. explore the relationship of this region with other world regions, particularly Europe and Asia;
6. have a better understanding of the historical migration patterns in the region;
7. comprehend the differentials in wealth and equality between various class and ethnic groups in the region; and
8. compare and contrast the different development policies that mark the different countries of this region.

Vocabulary Review

1. Land claims on the continent of Antarctica are regulated by the _____, which was signed in 1958 and includes 44 countries.

2. Throughout the outback of Australia one can find _____, where ranchers raise livestock on vast open ranges.

3. In Oceania most of the fishery resources are managed as _____, which affords all community members equal rights to access.

4. The entire region is vulnerable to global environmental changes. In particular, _____, which results in higher levels of ultraviolet radiation, has become a particular problem in Oceania.

5. The group of Aboriginal peoples who were forced into foster homes between 1928 and 1964 is often referred to as the _____.

6. Unlike the South Pacific Commission, which included the United States, France, and the United Kingdom, the _____, which was established in 1971 as an organization to promote regional cooperation, does not include any of the former colonial powers in the region.

7. In 1983, New Zealand and Australia built upon their supranational economic organization, the New Zealand–Australia Free Trade Agreement, by signing the _____, which set out to remove all tariffs and restrictions on trade between the two countries.

8. In the early 1900s, the Australian government set forth the _____, which was intended to maintain Australia's European "look."

9. The _____ emerged in Melanesia and is associated with the belief that a new age was coming when goods were brought to the region by foreigners.

10. In 1840 the British annexed New Zealand as part of the _____, which was signed by 40 Maori chiefs.

11. A(n) _____ is defined as an island that is made up of a coral reef.

12. Under European rule many new species were introduced to the region in a process some call _____. This process led to the endangerment or extinction of numerous indigenous species in the region.

13. One of the most unique features of Australia is the proliferation of _____, animals that give birth to premature offspring that must then be fed by nipples located in an outer pouch on the mother's body.

14. Based on the theory of _____, it is believed that larger islands will have a greater degree of biodiversity than smaller islands.

15. The most common form of tree in Australia is the _____, which is more commonly referred to as a gum tree.

16. _____ refers to the group of islands north of the equator.

17. Papua New Guinea and Fiji are just two islands that are part of _____, a region of islands found in the western Pacific.

18. The area in Australia known as the _____ is so named because it is underlain by the world's largest aquifer.

19. Even though income levels are low throughout much of the Pacific islands, the area is considered to be a place of _____ because people can rely on the rich supply of fish and coconuts to sustain a decent diet and support relatively high life expectancies.

Activities & Problems

1. Using Figure 11.7, answer the following questions.

   (a) What type of climate and vegetation would you find in Cairns? _____ and _____

   (b) What type of climate and vegetation would you find in Alice Springs? _____ and _____

   (c) What type of climate and vegetation would you find in Broome? _____ and _____

   (d) What type of climate and vegetation would you find in Sydney? _____ and _____

   (e) What type of climate and vegetation would you find in Brisbane? _____ and _____

   (f) What type of climate and vegetation would you find in Hobart? _____ and _____

2. Using Figure 11.3, answer the following:
   (a) Describe the physical landscape geography of Australia.

   _____

   _____

   _____

   _____

   _____

   _____

   (b) Describe the physical landscape geography of New Zealand.

   _____

   _____

_____

_____

_____

(c) Using the text and Figure 11.3, compare and contrast the physical landscape geographies of Australia and New Zealand.

_____

_____

_____

_____

_____

_____

(d) Which cities in Australia have the greatest populations? Where are they located? Why do you think those places have fostered such large populations relative to the rest of the country?

_____

_____

_____

_____

_____

_____

_____

3. Using Figure 11.13, answer the following questions.

(a) In what year did Cook make it to the eastern coast of Australia? _____

(b) Where did Cook's "third voyage" end? _____

(c) In what year did the British claim (and settle) Sydney? _____ Newcastle? _____ Hobart? _____ Darwin? _____ Perth? _____

(d) In what year did the Dutch claim Van Diemen's Land? _____ How far east did Abel Tasmen get on his voyage? _____

4. Using the subsection "Geographies of Indulgence, Desire, and Addiction: Uranium in Oceania," pages 550–551, answer the following questions:

(a) What is uranium and why did it become so important in the post-WWII period? Where can one find uranium today in this region?

_____

_____

_____

_____

_____

_____

(b) What is significant about nuclear testing in this region?

_____

_____

_____

_____

_____

(c) What have been some of the consequences of nuclear testing in the region?

_____

_____

_____

_____

_____

_____

_____

_____

_____

(d) What have some private organizations and local governments done to limit nuclear testing in the region?

_____

_____

_____

_____

_____

(e) What have been the results of the protests against nuclear testing in the region?

_____

_____

_____

_____

_____

_____

Review Exam

1. What are some of the main issues related to indigenous rights in Australia and New Zealand?

_____

_____

_____

_____

_____

2. Briefly examine the various economic practices associated with early colonialism in Australia and New Zealand and the implications for in-migration to the region.

_____

_____

_____

_____

3. What were the implications of the Treaty of Waitangi in New Zealand?

_____

_____

_____

_____

_____

4. How have the Pacific Islands been integrated into the world economy?

_____

_____

_____

_____

_____

_____

5. How were Australia's immigration policies modified in 1973? What was the impact of that change?

_____

_____

_____

_____

_____

_____

6. In what ways has the restructuring of global agriculture trade impacted New Zealand?

(a) There has been an increase in the export of traditional exports such as lamb meat.

(b) There has been an increase in the export of traditional exports such as wool.

(c) There has been an increase in the export of new exports such as wine and kiwi.

(d) There has been an increase in the export of new exports such as wool and wine.

7. Which of the following is true of ecological imperialism in the region?

(a) The dingo was first introduced by Europeans in the 19th century.

(b) The region's flightless birds were most vulnerable to the influx of rats, cats, dogs, and snakes brought by Europeans.

(c) During the colonial period Europeans limited the introduction of weeds, pests, and non-indigenous crops.

(d) The recent introduction of the cane toad to Australia has helped aid in reducing other non-indigenous species.

8. Which Australian policy restricted immigration to people from Northern Europe?

(a) White Australia Policy

(b) Treaty of Waitangi Policy

(c) Closer Economic Relations Policy

(d) Assimilation Policy

9. Which of the following was not a reason that decolonization of the Pacific Islands was slow to occur after World War II?

(a) The islands were of strategic significance to the United States and other allied nations.

(b) The islands were of economic significance to the colonial powers.

(c) The islands were of importance to the advancement of nuclear technology.

(d) The islands were important sources of migrant labor to the newly booming United States economy.

10. The ethnic Samoan diaspora is currently centered in which of the following countries?

    (a) United States   (b) France

    (c) United Kingdom   (d) Germany

11. Which of the following is not true of Indo-Fijian and ethnic Fijian relations in Fiji?

    (a) The conflict is a legacy of British colonial rule.

    (b) By 1960 Indo-Fijians almost outnumbered ethnic Fijians.

    (c) There have been high levels of intermarriage between Indo-Fijians and ethnic Fijians since 1990.

    (d) In 1999, an Indo-Fijian government was elected to head the government.

12. Which of the following countries does not lay claim to Antarctica?

    (a) United Kingdom   (b) Argentina

    (c) France   (d) Germany

13. Which of the following is one of the fastest growing industries in the Pacific Island region today?

    (a) tourism   (b) car manufacturing

    (c) textiles   (d) mining

14. Which of the following countries does not maintain any colonial or political ties with island nations in the Pacific?

    (a) United States   (b) Japan

    (c) United Kingdom   (d) France

15. Which of the following is true of Cook's journey to the Pacific?

    (a) He made three voyages to the area.

    (b) He was killed during a mutiny in Tonga.

    (c) He helped found Hobart.

    (d) He took prisoners and established Sydney in 1788.

For each of the following statements, write T for "True" or F for "False" on the space provided.

16. Micronesia means "dispersed islands." _____

17. A marsupial gives birth to a premature offspring. _____

18. The theory of island biogeography states that larger islands are likely to have higher levels of bio-diversity. _____

19. Eucalyptus is highly flammable and increases the potential for forest fires. _____

20. The demand for Australian wool was driven by the industrial revolution in the United States during the 19th century. _____

21. In the 1970s the economies of the region began to shift from Europe to Asia. _____

22. The Maori in New Zealand have much less power and recognition then Aborigines in Australia. _____

23. Southeastern Queensland is sometimes referred to as the "Gold Coast." _____

24. The climate of Oceania is marked by a high degree of seasonality. _____

25. In the 1920s both Australia and New Zealand made conscious efforts to develop local industry through the development of import substitution policies. _____

Key Map Items

Identify the following features on Map 11.1 (countries and cities) and Map 11.2 (landforms and waterways). On your political map, Map 11.1, label each city by placing a dot to mark the approximate location of that place. When mapping some of the landforms, you may need to create some sort of symbol system that portrays the landform, such as a ^^^^^^^^^^^^^ for a mountain range.

Countries	Major Cities	Landforms	Waterways
As listed in Chapter 11 of your textbook (Figure11.1)	• Adelaide • Alice Springs • Auckland • Brisbane • Cairns • Canberra • Darwin • Melbourne • Papeete • Perth • Port Moresby • Sydney • Wellington	• Ayres Rock • Blue Mountains • Eastern Highlands • Great Dividing Range • Great Victorian Desert • Interior Lowlands • MacDonnell Ranges • Southern Alps • Tasmania • Micronesia • Melanesia • Polynesia • Great Barrier Reef	• Timor Sea • Indian Ocean • Great Australian Bight • Arafura Sea • Timor Sea • Coral Sea • Tasman Sea • Bass Strait • Cook Strait

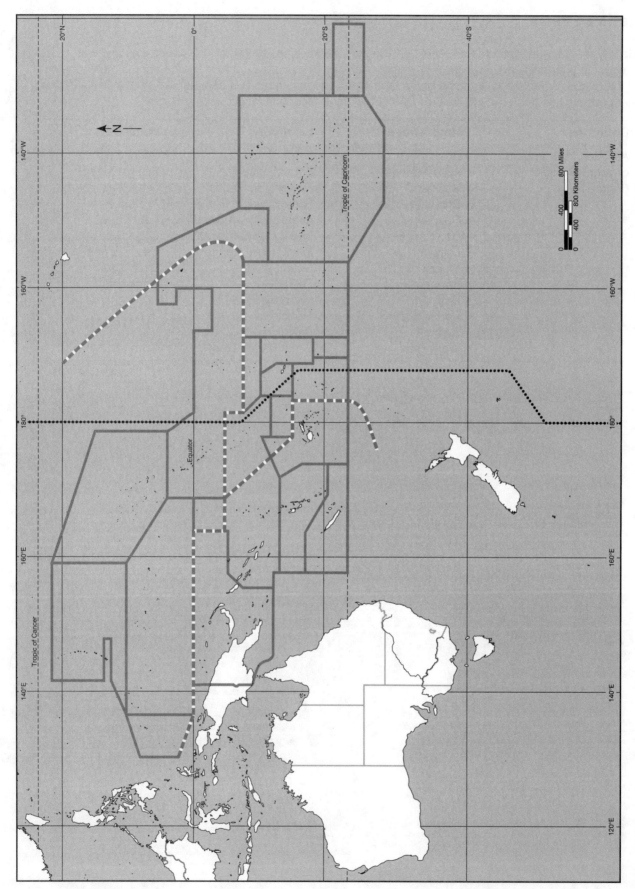

Map 11.1

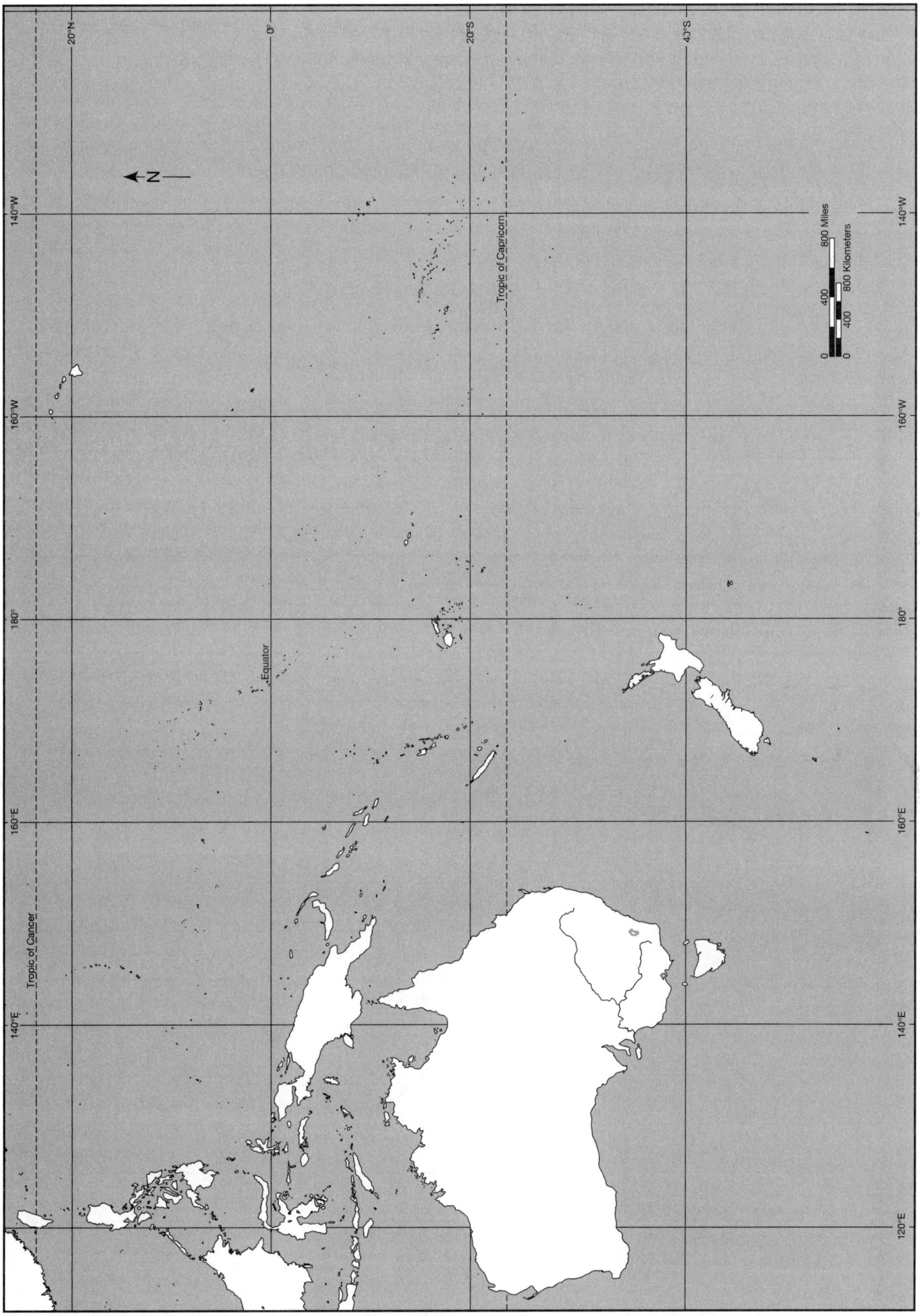
N
20°N
0°
20°S
40°S
140°W
160°W
180°
160°E
140°E
120°E
Tropic of Cancer
Equator
Tropic of Capricorn
0 400 800 Miles
0 400 800 Kilometers

Map 11.2

Mapping Human-Environment Relations

Goal: In this exercise, you are going to vegetation geography in relation to the geography of primary commodity production in Australia. You want to do a detailed reading of Chapter 11, particularly Figures 11.3, 11.7, 11.14, and 11.22.

1. Using the attached map and different colored pencils outline the climatic and vegetation geography of the region, providing a key for each region represented on the map.

2. Using points and outlines, highlight the wheat, sugarcane, sheep, cattle, and mining production in the country.

3. Add to the map, in text, key physiographic features that might have a direct impact on the primary commodity geography of this region (e.g., Great Dividing Range).

4. Where are the commodities you mapped developed relative to the map of vegetation and climate? Explain the relationships.

_____

_____

_____

_____

_____

_____

_____

5. How have humans had to adapt their environment to develop the primary industries discussed in Map 11.14? Explain.

_____

_____

_____

_____

_____

_____

_____

_____

_____

6. What international factors affected the spatial organization of primary commodity production in Australia? Explain.

_____

_____

_____

_____

_____

_____

_____

_____

_____

_____

7. Where are aboriginal lands located in relation to the various vegetation, climatic, and resource geographies you have already mapped? Explain.

_____

_____

_____

_____

_____

_____

_____

_____

_____

_____

_____

_____

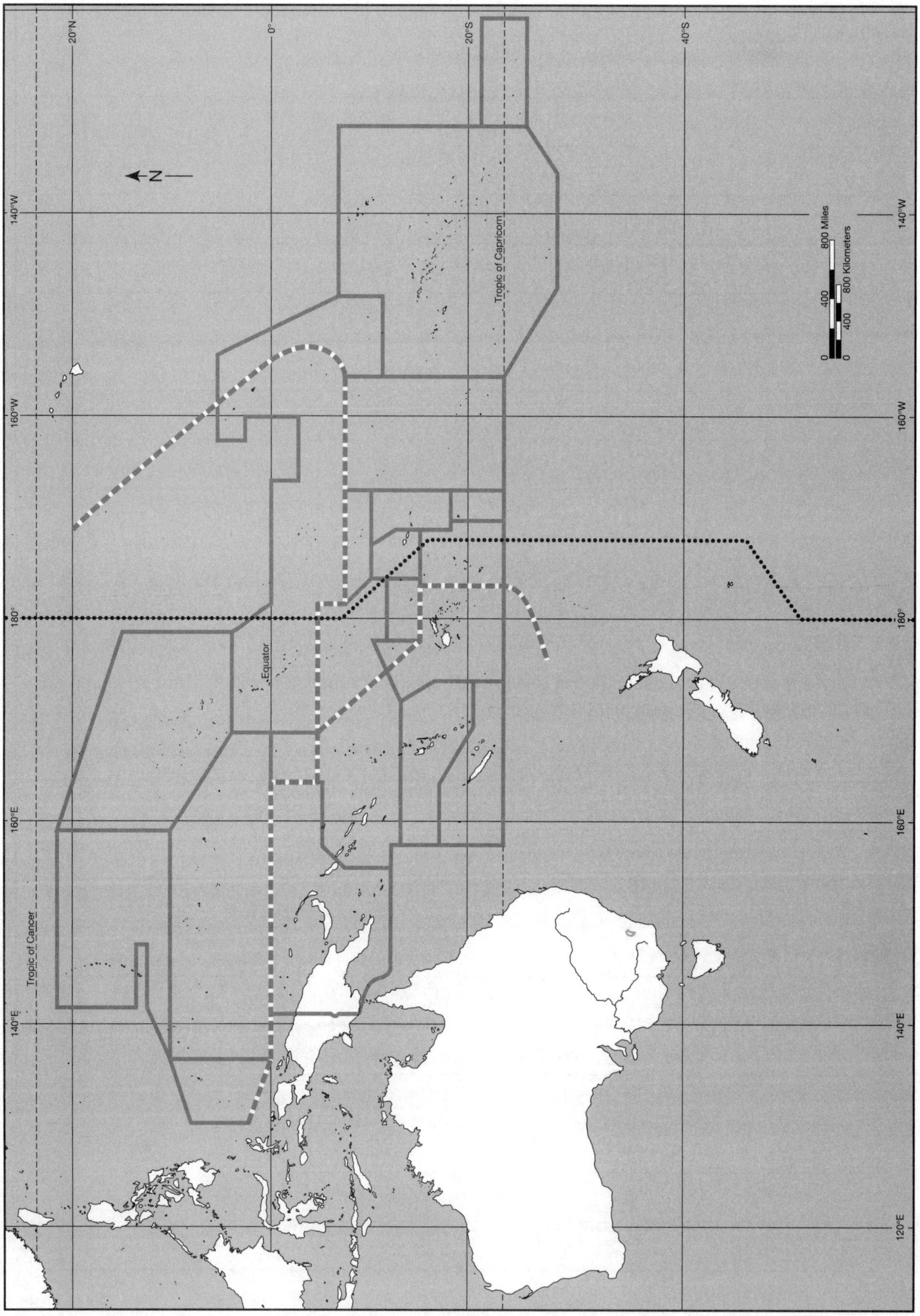

Map 11.3

CHAPTER 12

Future Regional Geographies

Learning Objectives

After reading and studying this chapter, you should be able to

1. distinguish between the different theories about the global future;
2. compare and contrast perspectives on globalization;
3. examine the growing geographic interconnections that have both brought people closer together and have fostered further inequities and divisions;
4. understand debates that have emerged around sustainability and regional development;
5. comprehend the changing cultural geographies brought on by increased interconnections between global processes and local practices;
6. discuss the prospects and future of the "state" and "global governance;"
7. examine the potential for new regional geographies; and
8. compare and contrast resource allocation at local, regional, and global scales.

Activities & Problems

1. Using Figure 12.13, discuss some of the examples of global patterns of invasive species and briefly discuss their impact.

2. Respond to the following based on your reading of Figure 12.5.

    (a) Which country in North America has the lowest index of income inequality?

    _____

    (b) Which country in Oceania has the highest index of income inequality?

    _____

(c) Overall, which region has a higher index of income inequality, Central America or the Caribbean? _____

3. Based on Figure 12.17, what can you say about world trade interconnections? What trends do you see? What regional disparities are apparent?

_____

_____

_____

_____

_____

_____

_____

_____

_____

_____

Review Exam

1. Briefly explain what is meant by the term the "short 20th century."

_____

_____

_____

_____

_____

_____

2. What is meant by the term "hyperglobalist"?

_____

_____

_____

_____

3. What are some of the limitations for development that are being faced by the marginalized regions of the world today?

_____

_____

_____

_____

_____

_____

4. What are four reasons Jane Jacobs gives for arguing that the United States may soon enter a "Dark Age"?

_____

_____

_____

_____

_____

_____

5. Why are experts predicting that 50% of the world's energy consumption will take place in the peripheral and semiperipheral regions by 2010 (i.e., what is driving this change in energy consumption)?

_____

_____

_____

_____

_____

6. What is Japan doing to reduce their dependence on energy resources?

_____

_____

_____

7. Why is it likely that there will continue to be environmental sustainability problems in peripheral and semiperipheral regions?

_____

_____

_____

_____

_____

_____

8. What is the foreseen impact of President Bush pulling the United States out of the 1997 Kyoto Agreement?

_____

_____

_____

_____

_____

9. What accounts for the "fault lines" that have appeared between previously compatible cultural groups in recent years?

_____

_____

_____

_____

_____

10. What are examples of growing global transnational governance?

_____

_____

_____

11. What is meant by the term "international regime"?

12. What are the possible reasons for pursuing regional integration?

13. What are some of the limitations to the sustainability of regional integration?

14. Why do the skeptics believe that we are moving toward regionalization not globalization?

_____

_____

15. Compare and contrast the possible "future scenarios" related to our environmental future. Which do you think seems most plausible? Explain your answer.

_____

_____

_____

_____

_____

_____

Appendix I: Answers

Preface

Answers to "Testing Your Map Knowledge"

1. latitude; longitude
2. prime meridian; equator
3. nominal; absolute
4. global positioning system
5. site and situation
6. physical attributes; a place relative to other places and human activity
7. cognitive maps
8. equal-area; outline
9. Remote sensing
10. c
11. (i) a, (ii) d
12. data capture
13. geodemographic research

Chapter 1: A World of Regions[1]

Vocabulary Review

1. region; places

2. gaining cost advantages from high-volume production

3. regionalism

4. human society that has adapted to environmental challenges

5. primary; secondary

6. the interdependence between the Old World and the New World

7. demographic transition

8. GDP; GNP

9. irredentism

10. commodity chains

11. technology system

12. Plate tectonics

13. Weather; climate

14. intertropical convergence zone; ITCZ

15. biome; biogeography

16. Global warming; greenhouse effect

17. world-system

18. Western Europe; North America; western Europe; Japan

---

[1]As noted above, answers to the Key Map Items and various Mapping Exercises that follow the Review Exams of each chapter are not included in the answer key. The responses to these questions will vary based on your own analysis.

19. urbanization; colonization

20. diaspora; nationalism

21. Hegemony

22. division of labor

23. neocolonialism

24. an analysis of social change that assesses the economic progress of industrial countries

25. World Bank; International Monetary Fund; IMF

26. Neoliberalism

27. transnational corporation

Activities & Problems

1. North America, most of Western Europe, East Asia (Japan and South Korea). Australia and New Zealand.

2. Answers will vary.

3. (a) divergent

   (b) subduction

   (c) subduction

   (d) subduction

   (e) subduction

4. See textbook, page 23, for answers.

5. As sinking air moves from high to low pressure cells some of that air is forced back toward the equator, forcing it to rise once again, creating a band of constant high pressure around the equator

6. Any three of the following: Democratic Republic of Congo, Central African Republic, Cameroon, Gabon, Republic of Congo, Angola, Zambia

7. Any three of the following: Burma, Laos, Vietnam, or Thailand

8. French in 1714: Europe; French in 1914: Western/Northern Africa, Madagascar, Southeast Asia, South America; Spanish in 1714: North and South America, Southeast Asia (the Philippines); Spanish in 1914: Europe; France was confined to Europe in 1714 and expanded colonial holdings over next two centuries. Spain, on the other hand, held colonies in both the Americas and Asia in 1714, but by 1914 had retreated back to Europe.

Review Exam

1. Regional geography combines elements of human and physical geography; regional geography focuses on understanding human–environment relations through a regional context; regional geography examines how places combine to construct regional identities. Human and physical geography have historically been systematic fields that focus on one aspect of human or physical experiences, such as cultural geography, fluvial geography, economic geography, etc., across space.

2. Places (and regions) exert influence on people's collective memories, physical well being, opportunities, and lifestyle choices.

3. Globalization creates interconnections between economic, environmental, political, and social processes. Currently, most people live in places that are connected to world markets for goods and services (e.g., finance) through the flows of international trade and cultural exchange.

4. "Development that meets the needs of the present without compromising the ability of future generations to meet their own needs."

5. GDP is an estimate of the total value of all materials, foodstuffs, goods, and services produced by a country in a particular year. GNP includes the net value of income from abroad, including profits and losses from overseas investments.

6. The demographic transition is, in part, the result in the changing economic base of a country/region/place from preindustrial to industrial. As this movement in time occurs, populations initially boom. As industrial culture emerges, birthrates decrease and the population appears to stabilize.

7. Some countries become stalled in a period of high-growth rates in population. These high growth rates are attributable to increases in medical technologies that have reduced infant mortality while cultural attitudes toward large families lag behind such innovations. The result is larger families and increased populations.

8. The experience of everyday routines in familiar settings allows people to derive a pool of shared meanings.

9. Five Phases: 1790–1840, early mechanization of water-power and steam engines; 1840–1890, exploitation of coal-powered steam engines, steel products, railroads, world shipping; 1890–1950, internal combustion engine, oil and plastics, electrical and heavy engineering; aircraft, radio and telecommunications; 1950–1990, exploitation of nuclear power, aerospace industries, electronics and petrochemicals, development of limited-access highways and global air routes; 1990–present, exploitation of solar energy, robotics, microelectronics, advanced materials productions, and advanced telecommunications.

10. Spatial justice is a term used to represent the fairness of geographical variations in people's levels of affluence and well-being, given people's needs and their contributions to the production of wealth and social well-being.

11. Land masses have moved relative to each other and across Earth's surface over millions of years. He determined this by showing how fossils from Africa and South America were similar and the fit of South America and Africa as continents along their coastlines.

12. Biodiversity: differences in the types and numbers of species in different regions of the world. Biodiversity is usually related to temperature. Since a large portion of the world exists in the tropics, we find high levels of biodiversity in those regions.

13. 1800: Core of the world economy was in western Europe. The semiperiphery was in North America. Large parts of Asia and eastern Europe were peripheral, as was a small part of southern Africa and South America. The rest of the world was largely external to the world system. 2000: There were three major core regions in North America, western and northern Europe, and Japan. The semiperiphery was made up of large parts of Asia, eastern Europe, South and Central America and Oceania. Africa and Southwestern Asia as well as large parts of Southeast Asia were in the periphery, as were parts of South and Central America.

14. The modern core of the world-system, which includes North America, western/northern Europe, and Japan.

15. (1) Division of labor enabled increased productivity; (2) standardization of time, space, measure, value, and money afforded more predictability and consistency in commerce and manufacturing; (3) emergence of distinct national identities and powerful states; (4) control and commodification of nature; (5) development of physical infrastructures, improving movements of goods, peoples, and ideas.

16. Nation is a term that denotes a people that recognize a common identity, but they may not reside in one common geographical area. Nation–state is an ideal form of a statehood under which everyone has a common nationality. Nationalism is the feeling of belonging to a nation as well as the belief that a nation has a right to self-determination. Nationalism plays the role of fostering the desire to construct nation–states.

17. The development of transportation networks was vital to the development of core economies because they opened up numerous regions to agricultural and industrial exploitation. The mechanization of both agricultural and industrial production facilitated this development.

18. International division of labor refers to the process by which different people, regions, and places specialize in particular economic activities. Under colonization, colonial powers organized particular economic activities in each colonial holding based on local comparative advantages.

19. Between 1790 and 2000 Indian lands dramatically decreased as a total proportion of the continent's total land area. Indian lands in 1790 consisted of almost all of continental United States excepting the eastern seaboard and parts of Texas and Louisiana. Over the next 100 years, Indian lands shrank as a total percentage of the land area of United States as they were moved west. By 2000, there are only a few recognized Indian lands that can be discerned when projecting these places on a national scale map. The largest Indian land holdings are in the U.S. Southwest.

20. Colonization facilitated the movement of new foodstuffs globally. The potato, which was first domesticated in the Americas, was transplanted to Europe where it became a staple crop in a number of countries, including Ireland.

21. Rostow argued that all countries go through five stages of economic growth from "traditional society" to "high mass consumption." He argued that this was a natural trajectory and that countries only need follow the path toward diversified, industrial market economies to be on the right path toward development. Dependency theorists argue, however, that development does not take place in a black box and that global interconnections relegate certain peoples and places to secondary and tertiary positions within the global economy. Development, they argue, also created underdevelopment by taking the surplus out of the periphery to be exploited in the core. Feminists also critiqued traditional development theories, arguing that these models ignored the gendered division of labor and the role of women in economic development. Moreover, they argued that women experience development different from men.

22. Tropical climates can be found between the tropic of Cancer and the tropic of Capricorn on all continents. Dry and semiarid desert and steppe climates are also located throughout the tropics and also can be found in the midlatitudes. Mesothermal climates are found in parts of the southern United States, western Europe, east, southeast, and south Asia, South America, and the southernmost parts of Africa and Australia and New Zealand. Microthermal climates are located in the midlatitudes and high latitude regions of the Northern Hemisphere. Polar climates are found near both poles, while highland climates are most discernable at the global scale in the Himilayas, Rocky Mountains, Andes Mountains, and Ethiopian highlands.

23. In the tropical regions one would find tropical rainforests and savanna. In dry arid and semiarid regions there are mostly deserts (less than 10 inches of rain per year) and steppes. Mesothermal climates produce a variety of subclimates that sustain various forest lands and different types of vegetation. Polar and highland climates support little to no vegetation and, at best, are home to grasslands, and shrubs.

24. c	25. b	26. d	27. a
28. b	29. c	30. d	31. d
32. d	33. b	34. a	35. b
36. c	37. d	38. d	

Chapter 2: Europe

Vocabulary Review

1. northwestern uplands; north European lowlands; central plateaus; Alpine system

2. fjords

3. polder

4. acid rain

5. buffer zone; satellite

6. state socialist

7. balkanization

8. enclaves

9. ethnic cleansing

10. xenophobia

11. disinvestment; deindustrialization

12. Golden Triangle

13. fascism

14. municipal housing

15. Pastoralism

16. tundra

17. steppe

18. latifundia; minifundia

19. entrepot

Activities & Problems

1. (a) Northwestern Uplands

   (b) North European Lowlands

   (c) Central Plateau

   (d) Alpine System

2. Turkish; Algerian

3. Portugal; Spain; Italy; Yugoslavia; Greece; Ireland; Finland

4. Answers may vary. Some sample responses include: neocolonial connections between former colonies and European colonial powers; uneven development within Europe, whereby southern Europe has developed at a slower rate of growth than northern Europe; increased fluidity of borders because of the development of the European Union.

5. WWI and WWII both impacted birth and mortality rates, meaning that there are deficits in both men and women who were in Germany during those wars. A baby boom ensued after WWII, and then fertility rates slowed in the 1970s to the present day.

Review Exam

1. Trade with non-European countries allowed the development and refinement of several key technologies that expanded European interest in global travel. The discovery of gold and silver prompted the Spanish to further expand their trading networks and empire into both the Atlantic and Pacific regions.

2. A national economy in which all aspects of production and distribution are centrally controlled by government agencies

3. EU policy on agriculture had the effect of stabilizing agricultural prices and subsidizing farmers' incomes. These policies have led to a general trend away from mixed farming toward monocropping. Farm modernization has had some detrimental impacts, including increased exploitation of moorlands, woodlands, wetlands, and hedgerows. Additionally, there have been farm surpluses resulting from subsidization of farming.

4. Geographic situation provides access to southern and central Europe; home to former colonial power capital cities; it includes the industrial heartland of western Europe; and concentrated market of skilled labor and affluent consumers.

5. An emerging secondary core in Europe that stretches through a number of distinct physical geographic and cultural landscapes. There is a high rate of urban development, benefiting from deindustrialization in the Golden Triangle region that has pushed industry into the Southern Crescent, creating flexible production regions that allow for local areas to develop their own industrial culture.

6. Most Muslims in Europe can be found in the industrial cities of western Europe. France's Muslim population is predominantly from Algeria, Morocco, and Tunisia. Germany's Muslim population is mostly Turkish, while the United Kingdom's Muslim population is from East Africa and South Asia. These patterns partially reflect former colonial relations, particularly in the case of France and UK, while all modern-day migrants are driven largely by the post-WW II economic boom. Many Muslims came to fill jobs in the industrial sectors of western Europe, jobs that could not be filled by Europeans themselves.

7. a	8. d	9. b	10. d
11. c	12. b	13. b	14. a
15. d	16. a	17. F	18. T
19. T	20. F	21. F	22. F
23. T	24. T	25. T	26. T

Chapter 3: The Russian Federation, Central Asia, and the Transcaucaus

Vocabulary Review

1. tundra

2. taiga, tundra

3. chernozem; Siberian

4. unitary; federalist

5. mikrorayon

6. civil society

7. permafrost

8. state socialism

9. dry farming

10. salinization

11. territorial production complexes

12. Silk Road

13. irredentist; secessionist

14. iron curtain

15. Ural Mountains

16. Black Sea

17. oblasts; okrugs

18. apparatchiks

19. Commonwealth of Independent States

20. Near Abroad

Activities & Problems

1. (a) 16th

   (b) 17th

   (c) 17th

   (d) 18th

2. (a) Estonia, Latvia, Lithuania

   (b) Poland, Hungary, Romania

   (c) 1945–1955

   (d) 1946–1961

   (e) Poland, Czechoslovakia, Germany

3. West Siberian Plain: boggy, difficult terrain; Central Siberian Plain: uplifted rock shields with hilly uplands and deep river gorges; Russian Plain: good farmland, relatively flat.

4. Tundra: sparse vegetation with lots of rocky soil; Taiga: boreal coniferous forest; Steppe: tall, luxurious feather-grasslands with rich chernozem soil from high levels of decay.

5. Decollectivization has led to the breaking up of large-scale collective farms into small-privatized plots. These plots are only a few hectares and now break up the landscape into a diverse pastiche of different crops.

Review Exam

1. Extensive plains interrupted by the Ural Mountains. A mountain rim creates a physiographic border that stretches from the Iranian border to the Sea of Okhotsk. The plains regions are bordered to the south by the Central Asian Plateau that extends from Iran to the Balkans. The northernmost part of the region is marked by permafrost.

2. Stalin withdrew from the capitalist world economy and relied on vast Soviet resources to develop a highly centralized, bureaucratic, industrial state. Stalin pushed for the collectivization of agricultural production, and capital for creating industry was to be extracted from agriculture. Stalin relocated peasants onto collective farms. In order to put the plans into action, Stalin relied on high levels of repression and exploitation. Despite the high levels of repression, the Soviet economy boomed under Stalin, growing at 10% per annum. There were severe environmental impacts from rapid industrial development as well, since the system was largely unregulated.

3. Perestroiyka: rapid economic and governmental reforms and direct democratic participation; glasnost: independent elections at the local and national levels that gave way to removal of restrictions that had been placed on the legal formation of national identity under Stalin.

4. Decollectivization has been slow throughout Russia. In places where it has been successful, new crops are now planted, and crops are more spatially diverse. Villages remain largely intact, while some people have chosen to relocate themselves closer to their crops.

5. The rapidity of economic reform left a power vacuum in the market, which has been replaced by organized crime. This was possible because under the Soviet system there were no laws regulating business behavior, civil society, banking, or accounting, and there were no procedures for declaring bankruptcy.

6. a	7. b	8. d	9. c
10. b	11. a	12. a	13. b

14. a	15. a	16. F	17. T
18. T	19. F	20. F	21. T
22. T	23. F	24. F	25. F

Chapter 4: Middle East and North Africa

Vocabulary Review

1. aridity
2. oasis
3. without irrigation
4. afforestation
5. Ottomans
6. Balfour Declaration; Palestine
7. informal economic activities
8. import substitution; nationalization
9. submission; Muslim
10. Islamism; anticolonial/anti-imperialist movements
11. jihad
12. kinship
13. subsistence activity that involves breeding and herding animals to satisfy human needs for food, shelter, and clothing; transhumance
14. chador; veil
15. guest workers
16. hajj
17. intifada
18. Petrodollars
19. Zionism
20. internally displaced person

Activities & Problems

1. (a) French
   (b) British
   (c) Italian
   (d) British/Ottoman
   (e) French
   (f) British/Ottoman
2. (a) Saudi Arabia, Iran
   (b) Tunisia, Turkey, Israel, Bahrain
   (c) Algeria
   (d) Turkey, Israel

3. The migrations are into the oil-producing states, such as Saudi Arabia and Libya. There are also migrations for political reasons, such as out of western Sahara. Migrations began in the 1950s and were largely prompted by the development of oil economies in the Gulf States.

4. Israel in 1949 consisted of the older territory of Palestine except for the West Bank and the Gaza Strip. In 1967, Israel occupied Gaza, West Bank, Sinai Peninsula, and Golan Heights. In 1996, Israel remained in all the territories taken in 1967 except Sinai. Today, Gaza is under Palestinian control, as are parts of the West Bank.

5. Answers will vary.

Review Exam

1. A majority of the region is marked by dry arid/semiarid climates, although there are also midlatitude climates found along the north coast of Africa and along the Mediterranean coast of Israel, Lebanon, Syria, Turkey. Midlatitude climates structure into the Zagros Mountains of Iran. In southern Sudan there are both tropical climates along the border with the Central African Republic and Uganda and highland climates along the border of Ethiopia.

2. Plant and animal domestication facilitated the movement to agriculturally based minisystems during the Stone Age in the Fertile Crescent. The production of foods eventually led to the development of village-based economies. Populations slowly increased as food surpluses were produced, and social organization changed as people moved from relatively egalitarian nomadic groups to villages based on kinship networks. Leisure time increased and specialization in nonagricultural crafts was possible, giving way to new technological and social innovations. The beginning of more complex barters and trade systems developed.

3. Transhumance means the movement of herds according to seasonal rhythms. Transhumance differs from sedentary life because people move more regularly and are dependent on the collection of wild grains and on the occasional slaughter of their herds for food, clothing, and shelter. In some cases, women would be in charge of planting grains during the spring seasons.

4. Steps have been taken to reforest through afforestation programs throughout the region. Some countries are directly confronting desertification through the planting of greenbelts.

5. The rise of OPEC has defined the modern-day tensions over the much-needed commodity of oil, central to the industrialization of the global economy, particularly for rapidly industrializing nations. Despite being an economic commodity, oil is also a political commodity and has played an important role in regional and global politics throughout the latter half of the twentieth century and into the beginning of the twenty-first century. OPEC, for example, nationalized the oil industries in their countries and raised prices as a way to challenge western economic domination globally.

6. d	7. c	8. d	9. b
10. c	11. a	12. a	13. d
14. a	15. c	16. T	17. T
18. T	19. F	20. T	21. F
22. T	23. F	24. T	25. F

Chapter 5: Sub-Saharan Africa

Vocabulary Review

1. harmattan

2. domestication

3. Shifting cultivation

4. scramble for Africa; Berlin Conference

5. apartheid

6. homelands

7. circular migration

8. feminization of poverty

9. gender and development

10. microfinance; social capital

11. Horn of Africa

12. Slash and burn

13. Afrikaans

14. Middle Passage

15. hut; indirect

16. Royal Geographical Society

17. Millennium Development Goals

Activities & Problems

1. (a) salt, gold

   (b) salt, gold

   (c) some combination of oil, copper, diamonds, iron ore

   (d) salt, iron ore, uranium

   (e) some combination of diamonds, copper, uranium

2. (a) 1895; 1908; 1899; 1908

   (b) German East Africa; German Southwest Africa; Cameroon; Togo

   (c) Angola; Mozambique; British

   (d) Portugese; Belgian; British; French; German

3. In both cases contraceptive use increases with level of education. Overall, however, contraceptive use is higher in all cases for Botswana. This may be the case because in Nigeria there is a much higher level of women with no education.

4. Throughout the central region there is tropical rainforest indicating high levels of rainfall. The tropical rainforest does not make it to the east coast, probably due to the mountains of the Great Rift Valley. Just north and south of the tropical climate are bands of savanna, indicating less rain. Continuing north and south there are deserts and then finally Mediterranean climate regions. Along the Great Rift Valley there are also highland climatic regions.

5. The regions immediately surrounding the Sahara demonstrate moderate risk of desertification, as do good portions of the Horn of Africa. There are also patches of acute risk in the southern-most portions of the Sahel. Southern Africa is also witnessing moderate as well as acute risks for desertification.

6. Answers will vary.

Review Exam

1. There are five major basins located across the continent in the western, central, and southern regions. Mountain ranges can be found across the northern region known as the Maghreb, the Ethiopian Highlands in the Horn of Africa, and bordering the Great Rift Valley. Mountain ranges also frame the southern part of the continent on both the east and west coasts. Several major rivers and lakes have left marks on the landscape, creating the Nile and Congo valleys, for example.

2. According to all four maps, every country (except Somalia and Djibouti, where no data are available) has already reported AIDS cases. Over time, the southern and central regions of the continent have seen higher percentages of adults with AIDS.

3. This region is called the "cradle of mankind" because there have been the earliest signs of the development of human beings in this region. Indications of early ancestors of humans were found to be from 3.7 million years ago.

4. During the height of the Cold War, many countries in Africa were experiencing independence movements and transitions to postcolonial states. Leaders in various countries were exposed to socialism as well as communism. Both the Soviet and American blocs supported various factions as tensions mounted over control of the newly formed states. Examples include the civil violence in Angola.

5. The colonial period left many countries in the region underdeveloped politically. Politically, weak local social infrastructures have meant that numerous factions have been vying for control, and many countries are still marked by civil unrest and political instability.

6. c	7. a	8. d	9. a
10. c	11. a	12. d	13. c
14. a	15. d	16. T	17. T
18. F	19. F	20. T	21. F
22. T	23. T	24. T	25. T

Chapter 6: North America

Vocabulary Review

1. intermontane
2. europeanization
3. americanization
4. indentured servants
5. staple economy
6. hate crimes
7. Assimilation
8. multiculturalism
9. Internal migration
10. suburbanization
11. deindustrialization
12. creative destruction
13. farm crisis
14. Acid rain
15. Superfund sites
16. megalopolis; Main Street
17. Gentrification
18. ice, snow, and permafrost
19. federal states
20. tundra

Activities & Problems

1. Pacific mountains and valleys, intermontaine basins and plateaus, Rocky Mountains, Great Plains, interior lowlands, Appalachian highlands, Piedmont, Gulf-Atlantic coastal plain

2. (a) Brazil, China, United States, Zimbabwe

   (b) China, Brazil, India, United States

   (c) Africa and the Middle East

   (d) Zimbabwe, Brazil, Pakistan

3. (a) Between 1961 and 2006, there have been several shifts in labor force production in Canada including: (1) a decrease in primary sector work from 14% in 1961 to 6.1% in 1991 to 4.1% in 2006; (2) secondary production has been reduced more slightly, from 28.4% to 21.2% to 20.3%; and (3) the tertiary sector has continued to rise from 57.6% to 72.7% to 75.5%.

   (b) Canada has historically relied on primary and secondary production to maintain its economic standing. In recent years, a global shift in production has meant that Canada has also faced an increase in its service sector economy. It has not, to date, expanded as rapidly as the U.S. in the area of quaternary economic activities. The fact that Canada demonstrates a slight increase in secondary economic activities between 1991 and 2000 demonstrates the reliance on these industries and the important weight placed on them by the Canadian government.

4. (a) Large tracts of the South and the Southwest demonstrate high poverty only, while the region of megalopolis and California have large areas of high wealth only. Most of the country is home to areas of high poverty and wealth. Very few places are sites of high equality, although some of the places can be found scattered throughout the Midwest.

   (b) Answers may vary.

5. (a) The pattern demonstrates the three waves of migration to the U.S.: (1) 1820–1870, large migration stream from Northern and Western Europe; (2) 1870–1920, increase in Southern and Eastern European migrants accompanying Western and Northern Europeans; (3) 1970–present, Asian and Latin American migrants.

   (b) The most recent wave of migration is dominated by peoples from Latin America, particularly people from Mexico. Reforms on strict immigration legislation in 1980 fostered an increase in Mexican migrants to the U.S. where they now make up 7.3% of all foreign-born peoples in the country. Many come to the U.S. seeking work as migrant laborers or to work in factories.

   (c) The reduction in in-migration between 1930 and 1940 marks the high point in the U.S.'s strong anti-immigration stance, which was precipitated by a backlash against southern and eastern Europeans by the majority western and northern European populations that inhabited the U.S. at the time. Immigrants were seen as dangerous to U.S. values. The U.S. imposed race-based quotas to also reduce the reduction of Asians, particularly Chinese, migrants to the country.

Review Exam

1. Main reasons: vast natural resource base of land and minerals providing materials for industrialization; growing populations provided large and expanding market and cheap labor force; cultural/trading links with Europe provided business contacts; links to technological know-how and access to European markets.

2. Distributions of Native American subsistence practices were significantly related to the local climatic and vegetative geographies of the region. In the northernmost parts of the continent, Native Americans engaged in Arctic and sub-Arctic hunting and fishing. Moving south into the northeast part of the modern-day United States, Native Americans engaged primarily in woodland farming. In the southern part of the United States, they were also farmers. West into the Great Plains, hunting became the primary subsistence activity. In the Rocky Mountain region, Native Americans engaged in hunting and gathering, while along the coast fishing was the dominant mode of subsistence. Finally, in the Southwest and modern-day Mexico, Pueblos practiced sophisticated farming techniques that had been passed down for millennia.

3. The United States took on an aggressive policy in the Western Hemisphere in the late 19th and early to mid-20th centuries, a legacy of which still exists today. U.S. foreign policy in the region was guided, in part, by the 1823 Monroe Doctrine, under which President James Monroe decreed that European powers could no longer colonize American continents and that these same powers should not interfere with American-run governments. Theodore Roosevelt followed up on Monroe's doctrine arguing that the United States had a right to intervene in any Latin American or Caribbean country.

4. The Québec movement is based, in part, on a desire for self-determination among the French-speaking peoples of the province. Roman Catholicism and the French family were the core of this conservative region throughout most of its European history. These views led to the relative isolation of the province, which meant that the province was ill equipped to participate in the growing global U.S. economy. Québec was further marginalized by the economic growth of the western half of Canada and Toronto's development as a service sector core. Fear of losing hold of traditional values, local language, and socio-economic and political autonomy have resulted in the emergence of a secessionist movement in the region.

5. Cities became the basis of the "new economy" of manufacturing in the 19th century. This process continued into the 20th century, although development of urban core regions was differential, favoring the northern portions of the United States over the southern portion. Similarly, urbanization in Canada was tied to the border regions of manufacturing in the Great Lake regions and the Northeast of the United States

6. d	7. d	8. c	9. a
10. b	11. d	12. b	13. b
14. b	15. d	16. T	17. F
18. F	19. F	20. F	21. F
22. T	23. F	24. T	25. F

Chapter 7: Latin America and the Caribbean

Vocabulary Review

1. Altitudinal zonation
2. pristine myth
3. Treaty of Tordesillas
4. demographic collapse
5. haciendas
6. Plantations
7. banana republics
8. structural adjustment policies
9. Mulatto; zambo
10. urban primacy
11. Braceros
12. liberation theology
13. land reform
14. nontraditional agricultural exports
15. informal economy
16. Columbian Exchange
17. ecotourism

18. bioprospecting

19. Monroe Doctrine

20. offshore financial services

Activities & Problems

1. Tropical desert, highland climates, midlatitude seasonal climates, tropical savanna climates, mid-latitude seasonal climates, tropical savanna climates, tropical rainforest

2. (a) Answers include any three of the following: sugarcane, bananas, yams, maize, rice, poultry, pigs, and cattle.

   (b) Answers include any three of the following: coffee, maize, coca, tomatoes, melons, cut flowers, dairy products, and cattle.

   (c) Answers include any three of the following: wheat, barley, maize, potatoes, apples, sheep, guinea pigs, llamas, alpaca, and vicuña.

   (d) Answers include any three of the following: highland grains and potatoes, sheep, guinea pigs, llamas, alpacas, and vicuña.

3. (a) 1981–1988

   (b) 1973

   (c) 1982

   (d) 1983

   (e) 1954

4. (a) Portugese

   (b) Spanish

   (c) French

   (d) English

   (e) Dutch

   (f) Mayan

5. Mayan: Yucatan Peninsula and south into present-day Guatemala; Aztec: central and parts of southern Mexico stretching from the Pacific to the Caribbean; Inca: west coast of South America from the southern border of present-day Colombia to central Chile (dominated the Andes).

6. (a) South America's interior is dominated by the Amazon Basin, while the west coast of the continent consists of the Andes Mountains and a small coastal plain. North and south of the Amazon Basin are two highland ranges. South of the Brazilian Highlands is the area known as the Pampas, a relatively flat alluvial plain. Much of Central America is mountainous, and mountain chains surround central Mexico, leaving only a small coastal plain on the western side of the country. The Yucatan Peninsula is relatively flat. The Caribbean is a series of island chains that surround the Caribbean Sea.

   (b) Mountain ranges, such as the Andes, have resulted in dramatic differences in climate. These mountains, while situated in the tropics, still have snow pack. The Amazon Basin, on the other hand, is a flat region containing almost 20% of the world's freshwater. It is home to a combination of grasslands and rainforest. Volcanic activity is a staple of plate boundary countries, such as those in the Caribbean. Sharp peaks and tectonic activity mark many parts of the region. Because of the complex tectonic geographies of the region, the region is rich in numerous mineral resources.

   (c) Mineral wealth can be find throughout the old crystalline shield where the crust has folded, bringing older rocks to the surface of the earth. The Andes regions of Peru and Bolivia remain important mining regions. The region is also known for its rich oil, gas, and coal deposits, based, in part, on the much longer geologic history of the region.

(d) El Niño brings warmer, wetter winds to the coast of South America, a normally dry desert-like region. The increase in rains can cause flooding and high levels of rainfall. The result is sometimes global drought conditions. La Niña, on the other hand, creates rains in northeast Brazilian, drought in Mexico, and hurricane conditions in the Pacific.

(e) Biodiversity: high levels of biological uniqueness and variation. This particular region has high levels of biodiversity because of the tropical climates, which are able to sustain significant variations in biological organisms. Consistent temperatures and moisture facilitate the growth of myriad flora and fauna.

Review Exam

1. Theories for the decline of the Maya include overuse of lands that led to environmental degradation and social collapse as the state could no longer support large populations.

2. The Incas adapted to their environment by building terraces throughout the mountainous region. Terraces collected water and reduced soil erosion, while also providing a flat surface on which to plant. Moreover, the terraces reduced frosts by reducing downhill flows of cold air.

3. The stark differences between rich and poor have their roots in the colonial infrastructure and the hacienda system. Under that system, a minority was given control over land and resources. In the postcolonial period, oligarchies rose based on the hacienda system, and wealth continues to be concentrated in the hands of a few. The results have been rapid urbanization and poverty in both rural and urban areas as corrupt politics maintain high degrees of economic difference between rich and poor. In some cases, a middle class has emerged, but in many other cases, a two-tiered system remains.

4. Liberation theology comes out of the Catholic tradition and is focused on developing the economic, political, and social capabilities of the poor. It is a combination of the philosophies of Jesus Christ and Karl Marx, both of whom spoke about inequality and oppression. Priests developed grassroots self-help groups, or what are often called Christian-based communities. They often spoke out against repression and authoritarianism.

5. The geologic history of the region created considerable mineral wealth from precious and industrial metals (gold, silver, copper, tin, aluminum, etc.) to oil and natural gas. Mexico has been drilling oil since the 1890s and Venezuela was one of the founding members of the Organization of Petroleum Exporting Countries (OPEC) in 1961, but fossil fuel deposits extend throughout northern South America. Colombia and Brazil have not effectively exploited their petroleum reserves for political and environmental reasons.

6. b	7. c	8. c	9. c
10. a	11. b	12. c	13. a
14. d	15. c	16. T	17. F
18. T	19. F	20. T	21. T
22. T	23. T	24. F	25. F

Chapter 8: East Asia

Vocabulary Review

1. zaibatsu

2. Pacific Rim

3. Ring of Fire

4. treaty ports

5. Keiretsu

6. Asian Tigers

7. counter-urbanization

8. feng shui; geomancy

9. backwash effects

10. agglomeration

11. chaebol

12. massif

13. pinyin

Activities & Problems

1. (a) desert shrubs

   (b) grasslands

   (c) evergreen and mixed forest

   (d) broad-leaved forests and forest steppe

   (e) temperate grasslands

   (f) rainforest

2. (a) Main islands of Japan, Formosa (Taiwan), Korea (North and South), Sakhalin Island

   (b) Expanded empire to include Manchuria, a region whose northernmost extent is the Soviet border.

   (c) In 1933, Japan extended its sphere of influence into the Great China Plain and almost to Beijing. They also began to assault China through the port city of Tsingtao.

3. Population density demonstrates a significant decrease moving east to west. Japan, Korea, Taiwan, and the coastal plain of China have very high population densities, while the Tibetan Plateau, Mongolia, and the region of the Taklamakan Desert have much lower population densities. Resource allocation in the region would have to be toward areas of high density, including the urbanized parts of the region.

4. (a) United States

   (b) Australia, Cuba, Peru, Mexico

   (c) United Kingdom, France

5. (a) East Asia is divided into three distinct physiographic regions: the Tibetan Plateau, the central mountains and plateaus, and the continental margin. This landform geography has created a particular vegetation pattern. That pattern includes grasslands in the Tibetan Plateau and deserts north of the plateau. In the eastern half of China and in insular East Asia, there is a distinct north-south vegetative zonation. Manchuria is marked by mixed needle- and broad-leaved forests. Northern Japan, North Korea, the northernmost parts of South Korea, and the North China Plain are areas of broad-leaved forests and steppes. In the southern region surrounding the Yangtze River there are subtropical evergreens, and in the southernmost parts of China one finds tropical rainforests.

   (b) Tibetan Plateau: grazing; Mongolia: pastoralism; North China Plain: wheat and barley cultivation; southern China: rice and tropical fruits.

   (c) Much of the land in the flat plains must be used for producing staple crops, such as rice and wheat, and this puts more pressure on land and population density.

   (d) Answers will vary.

   (e) Answers will vary.

Review Exam

1. Unified China under an imperial system and created stability that eventually facilitated the development of the Silk Road.

2. The imperial city included the following: Buddhist temples and Shinto shrines, palaces and castles, and gardens. Under the Tokugawa, the urban geography was altered as severe taxes on the agricultural economy were used to support industrialization in urban areas. Cities became the base of an "economic miracle" as Japan industrialized rapidly and created sophisticated transportation systems and industrial infrastructures. Rural-to-urban migration increased as jobs were created in the industrial sectors of urban Japan.

3. Tensions exist in Tibet and in Xinjiang. Tibet was invaded by the communist Chinese regime in 1950 and has been under "occupation" ever since. Tibet is predominantly Buddhist, and the Han Chinese wanted to control religion in the region. Today, temples are closely monitored by Han Chinese/communist officials. In Xinjiang, religion is also a major factor. In this case, however, it is a suppressed Muslim population that has pushed for greater autonomy. Many of the people of this region are also nomads, who came under the jurisdiction of strict mobility policies under the communist regime.

4. The Great Leap Forward was a scheme to rapidly increase economic productivity in China in the late 1950s. Land was merged into communes with the goal that industrialization would occur throughout the country in what came to be known as "backyard furnaces." It had a dramatic impact on the landscape and on everyday life. Large-scale communes replaced historically smaller farming plots, and the historical pastiche of multiple crops was replaced by monocropping. There was no planning on consumption, however, and industrialization in the countryside failed. Bad weather coupled with planning failures led to famine conditions, and 20 to 30 million people died.

5. Under the liberalized policies of Deng, migration from rural to urban regions increased dramatically, putting pressure on many urban areas. In the 1990s, the percentage of the population living in urban areas doubled. To alleviate stress on China's largest cities, the government created a series of buffer cities to absorb the flow of migrants. One of the major consequences of the increase in rural-to-urban migration has been a loss of arable land to housing, roads, and factories.

6. c	7. d	8. a	9. d
10. b	11. a	12. b	13. c
14. d	15. c	16. F	17. F
18. T	19. T	20. F	21. T
22. T	23. T	24. T	25. F

Chapter 9: Southeast Asia

Vocabulary Review

1. monsoon

2. Wallace's line

3. sawah

4. spice islands

5. culture system

6. transmigration

7. Overseas Chinese

8. world city

9. overurbanization

10. green revolution

11. tsunami

12. the Association of Southeast Asian Nations; ASEAN

13. domino theory

14. import substitution

15. Free Trade Zones; Export Processing Zone

16. land bridge

17. Swidden

18. Ring of Fire

Activities & Problems

1. (a) 5–8 months

   (b) 1–4 months

   (c) 0 months

   (d) 9–12 months

   (e) 0 months

2. Vegetation varies from monsoon forests sporadically located across mainland Southeast Asia to rainforest, also across the mainland and dominating much of the insular areas of the region. Monsoon forests can also be found sporadically located across parts of the insular region. Much of the mainland, the Philippines, Java, and Sumatra are marked by "other land use." Other land use stands in for agricultural land production. The map thus represents not only land cover but also land use. The river cultures of the mainland are marked by high degrees of agricultural productivity. Java is nearly completely deforested as are most of the Philippines. One can find secondary forest growth in parts of the Philippines and Laos, indicating either afforestation projects or severe land degradation that led to abandonment and secondary growth.

3. (a) Indonesia

   (b) Philippines

   (c) Thailand

   (d) Myanmar

   (e) Myanmar

   (f) Dong Sung (Vietnam); 700 B.C.E.

4. (a) Laos, Vietnam, Cambodia

   (b) Burma, Malaysia, Singapore

   (c) Indonesia

   (d) Philippines

   (e) Philippines

   (f) 1570; 1898

5. Japan first invaded East Asia taking Korea, Manchuria. In 1941, Japan extended its holdings to central China, French Indochina, Mariana Islands, and Marshall Islands. In 1942, Japan extended its influence into all of Southeast Asia, including Indonesia, Burma, Thailand, and Papua.

Review Exam

1. The insular region is bombarded with rains from both monsoons, while the mainland is only significantly affected by one monsoon season.

2. The highest levels of deforestation have been witnessed in the mainland, the Philippines, and Java. All three areas have the highest population densities in the region. The regions that have witnessed the least are the lower population density areas, such as Borneo, the mountainous regions of Sumatra, Papua, and the northernmost parts of Burma.

3. The transmigration program was an action taken by the Indonesian government to move people from regions of high density to regions of lower density, specifically off Java and onto other major islands. Problems include lack of infrastructure for migrants, conflict between migrants and indigenous groups, destruction of forests for agricultural production, malaria, pests, and weed invasion with migrants.

4. Overseas Chinese played the role of entrepreneurs in the colonial economy. They established banks, insurance companies, and shipping and agricultural businesses.

5. The Middle East and East Asia have received the largest numbers of migrants from Southeast Asia. The region is also marked by relatively high levels of internal migration between ASEAN members. Internal migration is largely to Malaysia and Singapore. Women work as maids and domestics in the Middle East. Philippine men have joined the merchant marines. Thais and Filipinos also work in Hong Kong and Taiwan because they are paid less than local labor. Filipino women also work as maids in East Asia and North America.

6. c	7. a	8. b	9. d
10. c	11. b	12. a	13. b
14. d	15. a	16. F	17. T
18. T	19. F	20. T	21. T
22. F	23. F	24. T	25. T

Chapter 10: South Asia

Vocabulary Review

1. Raj
2. Mountain Rim
3. Mughal
4. Hinduism
5. Bollywood
6. Caste
7. Indus Plain
8. diaspora
9. orographic effect
10. Deccan Plateau

Activities & Problems

1. (a) Ladakh, Kashmir and Jammu, Belutshistan, Rajputana, Fatehpur, Nitzam's Dominions, Mysore, Bhutan, Carchar, Manipur

   (b) 1753–1805: annexation took place along the eastern coast and up through the Ganges and Brahmaputra river valleys. After 1805: further extended control into the central regions of the country, north into present-day Pakistan and east into the northeast frontier and Assam.

   (c) Afghanistan and Nepal

2. Hinduism: Most of India, north coast of Sri Lanka. Islam: Pakistan, Bangladesh, and pockets throughout India, most highly concentrated along the borders with Pakistan and Bangladesh. Buddhism: Sri Lanka, parts of Nepal, Bhutan, and parts of northern India near Jammu and Kashmir. Christianity: east of Bangladesh in India, pockets along west coast of India and a few pockets along east coast of India, as well. Sikhism: north of New Delhi in India along the Pakistan border.

3. (a) Afghanistan

   (b) Bangladesh

   (c) Afghanistan, Nepal, India

4. (a) India is surrounded by a coastal fringe that gives way quickly to a series of mountains. The interior of India is distinguished by the Deccan Plateau, which extends to the Ganges and Bhramaputra River valleys. Surrounding the Ganges River valley are plains that eventually meet up with the "mountain rim," which includes the Hindu Kush and Himalayan Mountains, and the Khasi and Chttagong hills.

   (b) 1. Plains

   2. Peninsular Highland

   3. Mountain Rim

   4. Coastal Fringe

   5. Mountain Rim

5. (a) The Soviet's believed that the instability in Afghanistan posed a serious security problem along its southern border and that insurgency had led the government to be less interested in acting as a satellite of the Soviet regime.

   (b) Civil war ensued after the Soviets pulled out in 1989. Civil war erupted between the *mujahideen* militias vying for control of the war-torn country.

Review Exam

1. The plains are dominated by three main river systems: Indus, Ganges, and Bhramaputra, flowing over a region of young sedimentary rock. The Ganges and Bhramaputra meet and form a giant delta region. These rivers provide the region with a steady, yet uneven, amount of snowmelt. Irrigation in the plains allows for large-scale agriculture and high population densities. The coastal fringe, on the other hand, consists of a mixed number of physical features. This is a relatively narrow stretch of land marked by marine erosion that has exposed the edge of the ancient shield of the Peninsular Highlands. Some parts of the coastal fringe have alluvial deltas from water that runs down the Western and Eastern Ghat Mountains. The land is relatively fertile and has supported population densities, which are not as high as the plains region.

2. Summer monsoons bring rain from the Indian Ocean into the South Asian region, dropping particularly heavy amounts of rain in Bangladesh and Pakistan and along the western coastal plain. In the winter, the winds switch and bring rain to the east coast of Sri Lanka and parts of the southern eastern coastal fringe. The impact has been to sustain large-scale wet rice agriculture (along with other export crops) in the rich delta plains of the Ganges and Bhramaputra rivers.

3. Colonial legacy, language, and religion are the factors that fuel tension in Sri Lanka. Tamils, Hindus that speak Tamil, were brought to the island under British rule to work the fields and the bureaucracy. They were given disproportionate access to resources. The Sinhalese majority, who are Buddhists, resented the Tamil incursion, and under independent elections in the postcolonial period took control of the government. Tamils have since fought for complete autonomy for the northern region of the country where most Tamils live.

4. Based on its industrial geography, India's development remains uneven. Areas of industrial growth have seen much higher levels of economic growth and increasing rates of standards of living for the growing wealthy and middle classes. As an example, one can look at the "silicon valley" of India around Bangalore and Hyderabad. Areas with initial advantages in industrial development

have received increased benefits, while areas of relatively low capitalist productivity have suffered, particularly rural areas.

5. Aryan peoples, who were nomadic pastoralists, migrated into the region between 1500 and 500 B.C.E. developing agriculture in the region of the Ganges River through a massive deforestation and irrigation effort. The spread of agricultural civilization spread throughout the region, with a majority of the region being farmed by small-scale farmers.

6. d	7. a	8. a	9. c
10. a	11. c	12. b	13. b
14. b	15. d	16. T	17. F
18. T	19. T	20. T	21. T
22. F	23. F	24. T	25. F

Chapter 11: Australia, New Zealand, and the South Pacific

Vocabulary Review

1. Antarctic Treaty
2. cattle stations
3. common property resources
4. ozone depletion
5. stolen generation
6. South Pacific Forum
7. Closer Economic Relations Agreement
8. White Australian policy
9. cargo cult
10. Treaty of Waitangi
11. British Commonwealth
12. ecological imperialism
13. marsupials
14. island biogeography
15. eucalyptus
16. Micronesia
17. Melanesia
18. Great Artesian Basin
19. subsistence affluence

Activities & Problems

1. (a) tropical, woodland
   (b) subtropical, shrub land
   (c) tropical, shrub land
   (d) temperate (rain all year), temperate forest
   (e) subtropical, rainforest
   (f) temperate (rain in winter), temperate forest

2. (a) Moving from east to west one finds a thin coastal plain, followed by mountain ranges, an interior lowland, then a raised plateau, with plains just to the south of the plateau and just west and north of the plateau.

   (b) High mountain ranges mark the center of both islands with narrow coastal plains.

   (c) Answers will vary.

   (d) Answers will vary.

3. (a) 1770

   (b) Hawaii

   (c) 1788; 1804; 1803; 1869; 1829

   (d) 1643; Fiji

4. (a) Uranium is an element found in nature that can be used effectively to create nuclear weapons. One can find uranium in northern Australia, which, as a country, contains almost 27% of the world's uranium.

   (b) It is significant because of the importance these islands have played in the advancement of nuclear technology globally. Uranium, and nuclear tests more generally, link this region to the global geopolitical landscape of the nuclear age.

   (c) Nuclear testing has created both political and environmental problems in the region. Environmentally, islands have been made uninhabitable and nuclear fallout has impacted surrounding populated islands. Politically, people have been removed from islands controlled by nuclear powers for testing, thus displacing their homes and cultural traditions.

   (d) Private organizations, such as Greenpeace, have tried to stop nuclear testing, while countries, such as New Zealand, have banned all nuclear-powered vessels from entering its ports. People have also rioted against the French in Tahiti.

   (e) The results have been the banning of nuclear testing in the region.

Review Exam

1. Develop a more multicultural political and cultural system in both countries; increase indigenous landholdings; economic and social status of indigenous peoples. These issues have been treated differently in the two countries. New Zealand has developed a multicultural policy, whereby indigenous peoples are recognized and their rights/concerns addressed. Australia, while granting certain concessions to indigenous peoples, still remains a place where indigenous peoples have less power and recognition.

2. Australia and New Zealand's economies were first based on wool production, although wheat farming also became important in southeastern Australia. The discovery of gold brought people in from across the globe, including Europe and China. Railroads and livestock yards facilitated the growth of the economy. And sugarcane was introduced into the northeastern tropical region of Australia. New Zealand's economic development mimicked Australia's.

3. The Treaty of Waitangi in New Zealand has been used by both British settlers and indigenous peoples to lay claim to land in the country. Despite apparent land right protections, Maori peoples were systematically disenfranchised by British settlers. In 1975, court cases supported the Treaty and established New Zealand as an officially bilingual country, although Maori experience is still much worse than those of settler ancestry.

4. The Pacific Islands were first integrated into the global economy as European traders became more interested in some of the crops grown and resources found in this region. WWII further brought the Pacific Islands into the global arena as many of these places became sites of military bases for the Japanese and the United States. During the Cold War, the United States maintained a strong interest in the region militarily. Today, some islands in the region remain dependencies of the United States, France, and Britain. Today, many of the countries are drawn into the global economy because of their dependence on imported goods, transfer payments, infrastructural improvements, and the dependence on tourism for local economies.

5. In 1973, Australia removed its racist immigration policies, replacing racial requirements for skill requirements. This prompted a wave of immigration from Asia, with large numbers of migrants from Vietnam, Hong Kong, and Philippines.

6. c	7. b	8. a	9. d
10. a	11. c	12. d	13. a
14. b	15. a	16. F	17. T
18. T	19. T	20. F	21. T
22. F	23. T	24. F	25. T

Chapter 12: Future Regional Geographies

Activities & Problems

1. Answers will vary.

2. (a) Canada

   (b) Papua New Guinea

   (c) Central America

3. Many countries globally trade with over 120 different countries, demonstrating high levels of global interconnection. Despite that trend, there are regional disparities. Sub-Saharan Africa, on the whole, has far fewer trading partners than does the former French colonies of Southeast Asia. Mongolia has relatively little external trading activity as well.

Review Exam

(These answers are suggestions; and the depth of responses may vary.)

1. This term is sometimes use be historians to discuss the development of the world's "triadic core" model based in the United States, Europe, and Japan. This period is marked by consolidation of this core between World War I and the collapse of the Soviet Union in 1991.

2. Hyperglobalists believe that the days of the nation-state are numbered. Economies are on their way, if they have not already gotten there, to being "denationalized." States will facilitate the interconnections in the global economy.

3. Margins faced include unprecedented levels of demographic, environmental, economic, and social stress. Disease, limited resources, and violence are manifest in these areas and inhibit positive social and economic change. Politically, these areas are marked by high levels of political corruption, further limiting the possibilities for sustained economic growth and social change.

4. The roots of the Dark Age include: (a) corporate immorality; (b) universities now serve employers and do not focus on intellectual activities of the mind; (c) scientific research is being bought by corporations; and (d) neoliberal policies are abandoning urban and regional development.

5. This change will be driven by industrialization targeting growing worldwide consumer demands in the developing world. Both the manufacturing of those goods and the consumption of them after production will be increased dramatically in peripheral and semiperipheral regions, which have much larger populations.

6. Japan has led the world in the reduction of the need for fossil fuels for cars and trucks. Moreover, they are considering using ceramics in exchange for high-cost rare metals that have high levels of heat resistance.

7. Simply put, there is less money in peripheral areas to cope with environmental problems. Industrial cleanup, for example, is expensive. Moreover, there is less awareness in regions with lower levels of education and in places where people neither have the time or the mechanisms for gaining access to information. In urban environments, cycles of poverty create barriers to coping with the ever-growing needs of the urban poor.

8. The result of the decision is that the United States will continue to marginalize concerns related to global warming and the emission of Greenhouse gases.

9. Release of pressure suppressed during the Cold War; the globalization of culture and the desire to revive local culture as a countermeasure to globalization; increased levels of poverty and degradation have led to increased levels of criminal activity in the margins. This has resulted in fault lines emerging between ethnic neighbors as people fight for resources and maintenance of their cultural identities.

10. Regional supranational and regional organizations, such as the IMF or ASEAN, signal a move toward global transnational governance, as does the movement of transnational social movements.

11. International regime refers to the development of transnational capital, social movements, and professional organizations. The current era is an "international political era," which is evinced by the growing interconnections between local social actors in numerous places.

12. Regional integration is relevant for strategic and security reasons. It is also necessary in the development of trading blocks that enable certain economic advantages. Regional integration also gives increased leverage in international affairs and politics.

13. Limitations include state desires to maintain autonomy and independence. Fear of loss of cultural identity also increases apprehensions about regional integration. As regional interdependence increases, already heterogeneous states face increasing awareness at the local level of independent identities, increasing the possibility for ethnic or cultural tensions and fault lines.

14. They believe globalization is a myth and that the world is reorganizing around new political blocks. They believe that the trend toward regionalization is in opposition to the trend toward globalization and see regionalization as a stronger force.

15. Answers will vary.

Appendix B

Population Reference Bureau 2006 World Population Data Sheet

POPULATION REFERENCE BUREAU

inform　empower　advance

www.prb.org

2006 World Population DATA SHEET

The World's 10 Largest Countries in Population

2006

Country	Population (millions)
China	1,311
India	1,122
United States	299
Indonesia	225
Brazil	187
Pakistan	166
Bangladesh	147
Russia	142
Nigeria	135
Japan	128

2050

Country	Population (millions)
India	1,628
China	1,437
United States	420
Nigeria	299
Pakistan	295
Indonesia	285
Brazil	260
Bangladesh	231
Dem. Rep. of Congo	183
Ethiopia	145

Countries With the Highest Share of Their Surface Area Protected (2006)

Country	Percent of surface area protected
Venezuela	63
China, Hong Kong SAR	51
Zambia	42
Liechtenstein	40
Brunei	38
Tanzania	38
Saudi Arabia	37
Dominican Republic	33
Colombia	32
Estonia	31
Guatemala	31
Belize	30
Botswana	30
Germany	30
Switzerland	29

Protected area as a percentage of a country's total surface area is an indicator for the UN Millennium Development Goals. Protected areas contribute to environmental sustainability in multiple ways by maintaining biodiversity, safeguarding genetic resources, preventing soils from eroding, and supporting local livelihoods. They can provide valuable social and economic benefits. However, designation of protected areas alone is not sufficient to ensure these benefits. Protected areas must be carefully selected and managed for conservation goals.

The Top 15 HIV/AIDS Prevalence Countries (2005)

Africa

Country	Percent of population
Swaziland	33.4
Botswana	24.1
Lesotho	23.2
Zimbabwe	20.1
Namibia	19.6
South Africa	18.8
Zambia	17.0
Mozambique	16.1
Malawi	11.8
Central African Rep.	10.7
Gabon	7.9
Côte d'Ivoire	7.1
Uganda	6.7
Tanzania	6.5
Kenya	6.1

Outside Africa

Country	Percent of population
Haiti	3.8
Bahamas	3.3
Trinidad and Tobago	2.6
Belize	2.5
Guyana	2.4
Suriname	1.9
Papua New Guinea	1.8
Cambodia	1.6
Barbados	1.5
Honduras	1.5
Jamaica	1.5
Thailand	1.4
Ukraine	1.4
Estonia	1.3
Myanmar	1.3

WORLD POPULATION HIGHLIGHTS

In Many—But Not All—Countries, Most Married Women Wish to Limit Childbearing to Two Children.

One very useful indicator of women's ability to limit their number of children—and of the prospect for future fertility decline—is their desire to cease childbearing. In Vietnam, 92 percent of women who had two living children said that they did not wish to have any more children. In Nigeria, by contrast, that figure was only 4 percent.

Percent of married reproductive-age women with two living children who do not want another child

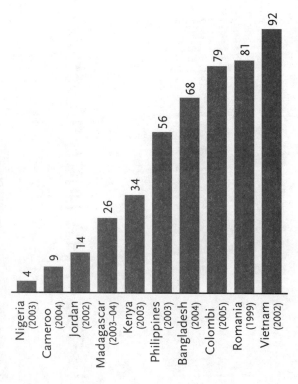

Sources: ORC Macro, MEASURE DHS STAT compiler, 2006; and Romanian Association of Public Health and Health Management and U.S. Centers for Disease Control and Prevention, *Reproductive Health Survey, Romania, 1999* (2001).

Net Migration Rates Vary Dramatically Around the World.

International migrants make up about 3 percent of the world's population. Economic conditions, social and political tensions, and historical traditions can influence a nation's level of migration. Net migration rates can mask offsetting trends (such as immigration of unskilled workers along with emigration of more-educated residents). Migration trends vary over time. For example, the Netherlands recently experienced a net outflow of people for the first time since the early 1980s.

Net migration rate per 1,000 population (2005)

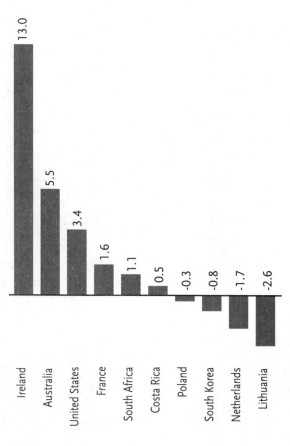

Source: PRB, *2006 World Population Data Sheet.*

For the graphic, "Many Governments Have Policies to Address Fertility Levels," please go to www.prb.org/wpds/map.pdf.

In Many Parts of the World, Rural Populations Still Lack Adequate Sanitation.

Worldwide, only 58 percent of the population has access to one of life's most fundamental needs: adequate or improved sanitation facilities. There are, however, wide regional and rural/urban disparities. In developing regions, only one-quarter to one-half of all rural residents have access to improved sanitation.

Percent of population with access to improved sanitation (2002)

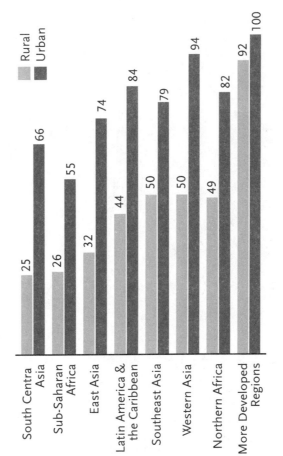

Source: UNICEF and WHO, Country, Regional, and Global Estimates on Water and Sanitation (2004).

In Some Poor Countries, More Than One-Fourth of Adolescent Girls Have Given Birth.

Fertility among women ages 15 to 19 presents a special concern, as these young women may lack the physical development and social support needed to carry a pregnancy to full term. Early childbearing can also curtail a young woman's education and reduce her potential earnings. Adolescents in the poorest countries—particularly in sub-Saharan Africa—are more likely to have given birth than adolescents in other countries.

Percent of women ages 15–19 who have given birth

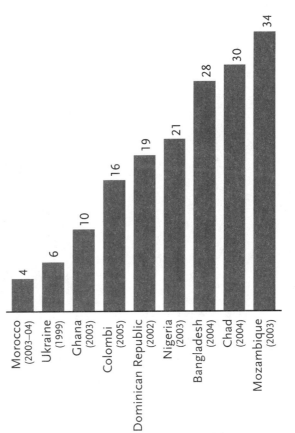

Sources: ORC Macro, MEASURE DHS STAT compiler, 2006; and Centers for Disease Control and Prevention and Macro International, *Reproductive, Maternal and Child Health in Eastern Europe and Eurasia: A Comparative Report* (2003).

Gains in Life Expectancy Since the 1950s Have Not Been Uniform.

In the early 1950s, life expectancy in China, Vietnam, Honduras, and Kenya was about 40 years—more than 30 years lower than in Sweden. Over the past half-century, China, Vietnam, and Honduras have each improved life expectancy by about 30 years—although they have taken different paths. For example, China experienced dramatic health improvements in the 1960s, while Vietnam's improvements became more pronounced in the 1970s and 1980s. As for Kenya, the HIV/AIDS crisis of the last 25 years has reversed much of the life expectancy gains of earlier decades.

Life expectancy at birth (years)

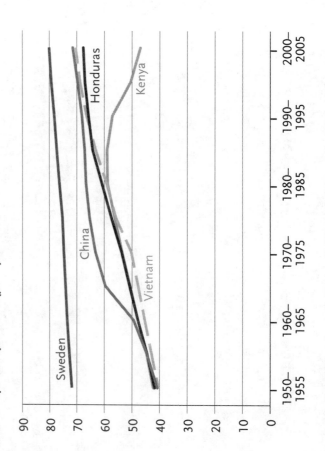

Source: United Nations Population Division, *World Population Prospects: The 2004 Revision.*

In Some Industrialized Countries, a Significant Share of the Population Lives in Economic Distress.

Almost everyone in the world's more developed countries lives well above the international poverty threshold of US$2 a day ($730 annually). That does not mean, however, that all persons in the industrialized world are economically well-off. Indeed, in many industrialized countries, more than one-tenth of residents have incomes below 50 percent of their country's median household income.

Percent of population in relative poverty (around 2000).

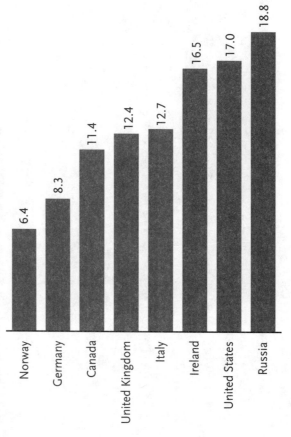

Source: Luxembourg Income Study, *LIS Key Figures* (www.lisproject.org, accessed July 23, 2006).

Demographic Data and Estimates for the Countries and Regions of the World

	Population mid-2006 (millions)	Births per 1,000 Population	Deaths per 1,000 Pop.	Rate of Natural Increase (percent)	Net Migration per 1,000 Pop.	Projected Population (millions) mid-2025	Projected Population (millions) mid-2050	Projected Pop. Change 2006–2050 (%)	Infant Mortality Rate[a]	Total Fertility Rate[b]	Percent of Pop. of Age <15	Percent of Pop. of Age 65+	Life Expectancy at Birth Total	Life Expectancy Males	Life Expectancy Females	% Urban	% of Pop. Ages 15–49 with HIV/AIDS 2003	% of Pop. Ages 15–49 with HIV/AIDS 2005	Percent of Married Women 15–49 Using Contraception All Methods	Modern Methods	Govt. View of the Birth Rate[c]	GNI PPP per Capita 2005	% Pop. Living Below US$2 per Day	Area of Country (Square Miles)	Population Density per Square Mile	Percent of Surface Area Protected 2006	Pop. With Access to Improved Sanitation (%) 2002 Urban	Rural
WORLD	6,555	21	9	1.2	0	7,940	9,243	41	52	2.7	29	7	67	65	69	48	1.0	1.0	61	54		9,190	53	51,789,601	127	12	81	37
MORE DEVELOPED	1,216	11	10	0.1	2	1,255	1,261	4	6	1.6	17	15	77	73	80	77	0.5	0.5	68	58		27,790	—	19,814,584	61	14	100	92
LESS DEVELOPED	5,339	23	8	1.5	-1	6,685	7,982	50	57	2.9	32	5	65	64	67	41	1.2	1.2	59	53		4,950	56	31,975,017	167	12	73	31
LESS DEVELOPED (Excl. China)	4,028	27	9	1.8	-1	5,209	6,545	63	61	3.4	35	5	63	62	65	42	1.5	1.6	50	42		4,410	59	28,278,917	142	12	74	32
AFRICA	924	38	15	2.3	-0	1,355	1,994	116	84	5.1	42	3	52	51	53	37	4.9	4.9	28	22		2,480	66	11,698,111	79	10	62	30
SUB-SAHARAN AFRICA	767	40	16	2.4	-0	1,151	1,749	128	90	5.5	44	3	48	47	49	34	6.2	6.1	22	15		1,970	—	9,379,573	82	11	55	26
NORTHERN AFRICA	198	26	6	2.0	-1	265	329	66	42	3.2	35	4	69	67	70	47	0.4	0.4	49	44		4,350	29	3,286,031	60	5	82	49
Algeria	33.5	21	4	1.7	-1	43.1	49.7	48	30	2.4	31	5	75	74	76	49	0.1	0.1	57	52	H	6,770	15	919,591	36	5	99	82
Egypt	75.4	27	6	2.1	-1	101.1	125.9	67	33	3.1	35	5	70	67	72	43	<0.1	<0.1	59	57	H	4,440	44	386,660	195	13	84	56
Libya	5.9	27	4	2.4	0	8.3	10.8	83	26	3.4	34	4	76	74	78	86	—	—	49	26	S	—	—	679,359	9	z	97	96
Morocco	31.7	21	6	1.6	-1	38.8	45.2	43	40	2.5	30	5	70	68	72	55	0.1	0.1	63	55	S	4,360	14	172,413	184	1	83	31
Sudan	41.2	36	9	2.6	-3	61.3	84.2	104	64	5.0	44	2	58	57	59	36	1.6	1.6	10	7	S	2,000	—	967,494	43	5	50	24
Tunisia	10.1	17	6	1.1	-1	11.6	12.2	20	21	2.0	27	7	73	71	75	65	0.1	0.1	63	53	S	7,900	7	63,170	160	2	90	62
Western Sahara	0.4	28	8	2.0	6	0.7	0.9	152	53	3.9	34	3	64	62	66	93	—	—	—	—		—	—	97,344	4	7	—	—
WESTERN AFRICA	271	43	17	2.6	-0	414	637	135	102	5.8	44	3	48	47	48	40	3.1	3.2	14	9		1,270	83	2,370,015	114	7	54	25
Benin	8.7	41	12	2.9	2	14.3	22.1	154	102	5.6	44	3	54	53	55	40	2.0	1.8	19	7	H	1,110	74	43,483	200	23	58	12
Burkina Faso	13.6	44	19	2.5	1	23.2	39.1	187	81	6.2	46	3	48	48	49	16	2.1	2.0	14	9	H	1,220	72	105,792	129	15	45	5
Cape Verde	0.5	30	5	2.5	-6	0.7	0.9	90	28	3.5	38	6	71	68	74	55	—	—	53	46	S	6,000	—	1,556	312	z	61	19
Côte d'Ivoire	19.7	39	14	2.5	4	27.1	36.1	84	104	5.1	41	3	51	49	53	47	7.0	7.1	15	7	H	1,490	49	124,502	158	16	61	23
Gambia	1.5	38	12	2.7	3	2.4	3.7	148	75	5.1	42	3	53	52	55	50	2.2	2.4	10	9	H	1,920	83	4,363	338	4	72	46
Ghana	22.6	33	10	2.3	-0	32.7	47.3	110	59	4.4	39	4	57	57	58	44	2.3	2.3	25	19	H	2,370	79	92,100	245	15	74	46
Guinea	9.8	41	13	2.8	-6	15.2	23.5	139	98	5.7	46	4	54	54	54	30	1.6	1.5	9	6	H	2,240	—	94,927	103	6	25	6
Guinea-Bissau	1.4	50	20	3.0	-1	2.4	4.4	225	116	7.1	53	3	45	44	46	48	3.8	3.8	8	4	H	700	—	13,946	97	7	57	23
Liberia	3.4	50	21	2.9	-8	5.8	10.7	217	142	6.8	47	2	43	41	44	45	—	—	—	—	H	—	—	43,000	78	13	49	7
Mali	13.9	50	18	3.2	-2	24.0	42.0	202	130	7.1	48	3	49	48	49	30	1.8	1.7	8	6	H	1,000	91	478,838	29	2	59	38
Mauritania	3.2	42	14	2.8	2	5.0	7.5	137	74	5.8	43	3	54	53	55	40	0.7	0.7	8	5	H	2,150	63	395,954	8	2	64	9
Niger	14.4	55	21	3.4	0	26.4	50.2	248	149	7.9	49	2	44	44	44	21	1.1	1.1	14	4	H	800	86	489,189	29	7	43	4
Nigeria	134.5	43	19	2.4	-0	199.5	298.8	122	100	5.9	43	3	44	43	44	44	3.7	3.9	12	8	H	1,040	92	356,668	377	6	48	30
Senegal	11.9	39	10	2.9	-2	17.3	23.1	94	61	5.3	44	3	56	55	58	45	0.9	0.9	12	10	H	1,770	63	75,954	157	11	70	34
Sierra Leone	5.7	46	23	2.3	-3	8.7	13.8	143	163	6.5	43	3	41	39	42	36	1.6	1.6	4	4	H	780	75	27,699	205	4	53	30
Togo	6.3	38	12	2.6	0	9.6	13.5	115	90	5.1	44	3	55	53	57	33	3.2	3.2	26	9	H	1,550	—	21,927	288	11	71	15

Column group headers: Demographic Data · Economy · Area & Density · Environment

NOTES

(—) Indicates data unavailable or inapplicable.

z Rounds to zero.

a Infant deaths per 1,000 live births. Rates shown with decimals indicate national statistics reported as completely registered, while those without are estimates from the sources cited on reverse. Rates shown in italics are based upon fewer than 50 annual infant deaths and, as a result, are subject to considerable yearly variability.

b Average number of children born to a woman during her lifetime.

c H=too high; S=satisfactory; L=too low.

d Special Administrative Region.

e The former Yugoslav Republic.

f Data are for the former Serbia and Montenegro.

* Data prior to 2000 are shown in italics.

Data prepared by PRB demographer Carl Haub.

Demographic Data and Estimates for the Countries and Regions of the World

	Population mid-2006 (millions)	Births per 1,000 Population	Deaths per 1,000 Pop.	Rate of Natural Increase (percent)	Net Migration per 1,000 Pop.	Projected Population (millions) mid-2025	Projected Population (millions) mid-2050	Projected Pop. Change 2006-2050 (%)	Infant Mortality Rate	Total Fertility Rate	Percent of Pop. of Age <15	Percent of Pop. of Age 65+	Life Expectancy at Birth Total	Life Expectancy Males	Life Expectancy Females	% Urban	% Pop. Ages 15-49 with HIV/AIDS 2003	% Pop. Ages 15-49 with HIV/AIDS 2005	Contraception All Methods	Contraception Modern Methods	Govt. View of the Birth Rate	GNI PPP per Capita 2005	% Pop. Living Below US$2 per Day	Area of Country (Square Miles)	Population Density per Square Mile	Percent of Surface Area Protected 2006	Sanitation Urban	Sanitation Rural
EASTERN AFRICA	284	41	16	2.4	-0	432	664	133	81	5.5	44	3	47	46	47	24	—	—	24	19		1,090	79	2,456,184	116	18	50	26
Burundi	7.8	46	18	2.7	7	14.0	25.8	229	106	6.8	46	3	45	44	45	9	3.3	3.3	16	10	H	640	88	10,745	729	6	47	35
Comoros	0.7	37	7	2.9	-2	1.0	1.5	118	59	4.9	43	3	64	62	66	33	<0.1	<0.1	26	19	H	2,000	—	861	803	3	38	15
Djibouti	0.8	31	12	1.9	-5	1.1	1.5	92	100	4.0	40	3	53	52	54	82	3.1	3.1	9	6	H	2,240	—	8,958	90	0	55	27
Eritrea	4.6	39	11	2.8	9	7.4	11.2	146	61	5.3	45	2	55	53	57	19	2.4	2.4	8	5	H	1,010	—	45,405	100	17	34	3
Ethiopia	74.8	39	15	2.4	0	107.8	144.7	94	77	5.4	44	3	49	48	50	15			15	14	H	1,000	78	426,371	175	17	19	4
Kenya	34.7	40	15	2.5	0	49.4	64.8	87	77	4.9	43	2	48	49	47	36	6.8	6.1	39	32	H	1,170	58	224,081	155	13	56	43
Madagascar	17.8	40	12	2.7	0	28.2	41.8	135	83	5.2	45	2	55	53	57	26	0.5	0.5	27	17	H	880	85	226,656	78	16	49	27
Malawi	12.8	44	18	2.6	-0	23.8	44.4	248	76	6.0	47	3	45	44	47	14		11.8	33	28	H	650	76	45,745	279	16	66	42
Mauritius	1.3	15	7	0.8	0	1.4	1.5	20	14.8	1.8	24	7	72	69	76	42	0.2	0.6	76	42	S	12,450	—	788	1,592	1	100	99
Mayotte	0.2	39	3	3.6	5	0.3	0.6	195	—	4.5	42	3	74	72	76	28			—	—			—	145	1,297	17		
Mozambique	19.9	41	20	2.0	-0	27.6	37.6	89	108	5.4	43	3	42	41	42	32	16.0	16.1	17	12	H	1,170	78	309,494	64	9	51	14
Reunion	0.8	19	5	1.4	2	1.0	1.1	34	7	2.4	27	7	77	72	80	89			70	—	S	—	—	969	818	3		
Rwanda	9.1	43	17	2.7	-0	13.8	20.6	128	86	6.1	47	2	47	46	48	17	3.8	3.1	17	10	H	1,320	84	10,170	890	8	56	38
Seychelles	0.1	18	8	1.0	-21	0.1	0.1	13	16	2.1	26	8	71	66	76	50			—	—	S	15,940	—	174	460	1	100	100
Somalia	8.9	46	17	2.9	5	14.9	25.5	188	119	6.9	45	3	48	46	50	34	0.9	0.9	8	1	S	730	90	246,201	36	14	47	14
Tanzania	37.9	42	17	2.5	-2	53.6	72.7	92	68	5.7	44	3	45	44	45	32	6.6	6.5	26	20	H	730	90	364,900	104	38	54	41
Uganda	27.7	47	16	3.1	-1	55.5	130.1	371	81	6.9	50	2	47	47	47	12	6.8	6.7	20	19	H	1,500	—	93,066	297	26	53	39
Zambia	11.9	41	23	1.9	-2	16.4	22.8	92	92	5.7	45	3	37	38	37	35	16.9	17.0	34	23	H	950	94	290,583	41	42	68	32
Zimbabwe	13.1	30	23	0.7	-1	14.4	15.8	21	61	3.6	41	3	37	38	37	34	22.1	20.1	54	50	H	1,940	83	150,873	87	15	69	51
MIDDLE AFRICA	116	44	16	2.8	-0	190	309	166	98	6.3	46	3	48	47	50	35	4.0	4.0	26	6		1,310		2,553,151	45	11	47	22
Angola	15.8	49	22	2.6	2	25.9	42.0	165	139	6.8	47	3	41	39	42	33	3.7	3.7	6	5	H	2,210	—	481,351	33	12	56	16
Cameroon	17.3	37	14	2.3	-0	24.3	32.3	87	74	4.9	43	3	51	50	52	53	5.5	5.4	26	13	H	2,150	51	183,568	94	9	63	33
Central African Republic	4.3	37	19	1.7	0	5.5	6.5	51	94	4.9	43	4	44	43	44	41	10.8	10.7	28	7	S	1,140	84	240,533	18	16	47	12
Chad	10.0	48	20	2.8	-3	17.2	31.5	214	101	6.7	47	3	44	43	45	24	3.4	3.5	11	2	S	1,470	—	495,753	20	9	30	0
Congo	3.7	40	14	2.6	-6	5.9	9.7	161	75	5.3	45	3	51	50	52	52	5.4	5.3	44	13	H	810	—	132,046	28	14	14	2
Congo, Dem. Rep. of	62.7	45	14	3.1	0	108.0	183.2	192	95	6.7	48	3	50	49	52	30	3.2	3.2	31	4	S	720	—	905,351	69	8	43	23
Equatorial Guinea	0.5	43	20	2.3	0	0.8	1.1	127	102	5.6	44	4	44	43	44	39	3.2	3.2	—	—	S	7,580	—	10,830	47	14	60	46
Gabon	1.4	33	13	2.0	-2	1.8	2.3	62	57	4.3	40	4	54	53	55	81	7.7	7.9	33	12	L	5,890	—	103,347	14	16	37	30
Sao Tome and Principe	0.2	34	9	2.5	-3	0.2	0.3	94	80	4.1	42	4	63	62	64	38	—	—	29	27	H	—	—	371	410		32	20
SOUTHERN AFRICA	54	24	19	0.5	1	55	55	3	55	2.9	33	5	46	44	48	50	19.3	19.5	54	53		11,460	36	1,032,730	52	14	84	41
Botswana	1.8	26	27	-0.1	-1	1.7	1.7	-6	56	3.1	38	3	34	35	33	54	24.0	24.1	40	39	H	10,250	50	224,606	8	30	57	25
Lesotho	1.8	28	25	0.3	-4	1.7	1.6	-11	91	3.5	39	5	36	35	36	13	23.7	23.2	37	35	H	3,410	56	11,718	154	z	61	32
Namibia	2.1	29	15	1.4	-0	2.5	3.1	49	44	3.9	43	3	47	47	47	33	19.5	19.6	44	43	S	7,910	56	318,259	6	15	66	14
South Africa	47.3	23	18	0.5	1	48.0	48.4	2	54	2.8	32	5	47	45	49	53	18.6	18.8	56	55	H	12,120	34	471,444	100	6	86	44
Swaziland	1.1	29	28	0.1	-1	1.0	0.8	-34	74	3.7	41	3	34	33	35	23	32.4	33.4	28	26	H	5,190	—	6,703	169	3	78	44
NORTHERN AMERICA	332	14	8	0.6	4	387	462	39	7	2.0	20	12	78	75	81	79	0.6	0.6	73	69		40,980		7,699,508	43	17	100	100
Canada	32.6	11	7	0.3	7	37.6	41.9	29	5.3	1.5	18	13	80	77	82	79	0.3	0.3	75	73	L	32,220	—	3,849,670	8	7	100	99
United States	299.1	14	8	0.6	3	349.4	419.9	40	6.7	2.0	20	12	78	75	80	79	0.6	0.6	73	68	S	41,950	—	3,717,796	80	23	100	100

Demographic Data and Estimates for the Countries and Regions of the World

	Population mid-2006 (millions)	Births per 1,000 Population	Deaths per 1,000 Pop.	Rate of Natural Increase (percent)	Net Migration per 1,000 Pop.	Projected Population (millions) mid-2025	Projected Population (millions) mid-2050	Projected Pop. Change 2006-2050 (%)	Infant Mortality Rate	Total Fertility Rate	Percent of Pop. of Age <15	Percent of Pop. of Age 65+	Life Exp. Total	Life Exp. Males	Life Exp. Females	% Urban	% Pop. 15-49 HIV/AIDS 2003	% Pop. 15-49 HIV/AIDS 2005	Contraception All Methods	Contraception Modern Methods	Govt. View of Birth Rate	GNI PPP per Capita 2005	% Pop. Below US$2 per Day	Area of Country (Sq Mi)	Pop. Density per Sq Mile	Percent Surface Area Protected 2006	Sanitation Urban	Sanitation Rural
LATIN AMERICA/CARIBBEAN	566	21	6	1.5	-1	700	797	41	26	2.5	30	6	72	69	75	76	0.5	0.5	71	63		7,950	24	7,946,684	71	18	84	44
CENTRAL AMERICA	149	24	5	1.9	-3	187	214	43	24	2.7	34	5	74	71	76	68	0.5	0.5	66	57		8,640	25	957,452	156	12	88	45
Belize	0.3	27	5	2.3	10	0.4	0.5	61	31	3.3	41	3	70	67	74	50	2.1	2.5	56	49	H	6,740	—	8,865	34	30	71	25
Costa Rica	4.3	17	4	1.3	1	5.6	6.3	48	10	1.9	28	6	79	77	81	59	0.3	0.3	80	72	S	9,680	8	19,730	217	23	89	97
El Salvador	7.0	26	6	2.0	-1	9.1	10.8	55	25	3.0	36	5	70	67	73	59	0.9	0.9	67	61	S	5,120	41	8,124	862	1	78	40
Guatemala	13.0	34	6	2.8	-4	20.0	27.9	115	35	4.4	43	4	67	63	71	39	0.9	0.9	43	34	H	4,410	32	42,042	310	31	72	52
Honduras	7.4	31	6	2.5	-2	10.7	14.7	100	30	3.9	42	3	71	67	74	47	1.5	1.5	62	51	H	2,900	44	43,278	170	20	89	52
Mexico	108.3	22	5	1.7	-4	129.4	139.0	28	21	2.4	32	5	75	73	78	75	0.3	0.3	68	59	S	10,030	20	756,062	143	9	90	39
Nicaragua	5.6	29	5	2.4	-4	7.7	9.4	67	36	3.3	40	3	69	66	70	59	0.2	0.2	69	66	H	3,650	80	50,193	112	18	78	51
Panama	3.3	22	5	1.7	0	4.2	5.0	52	19	2.7	30	6	75	73	78	62	0.9	0.9	—	—	S	7,310	17	29,158	113	25	89	51
CARIBBEAN	39	20	8	1.2	-3	48	55	41	40	2.6	29	8	69	67	71	64	1.5	1.6	61	57				90,653	433	8	82	51
Antigua and Barbuda	0.1	18	6	1.3	-6	0.1	0.1	0	21	2.3	28	5	71	69	74	37	—	—	—	—	S	11,700	—	170	406	1	93	94
Bahamas	0.3	19	9	1.0	-2	0.3	0.3	7	12.7	2.3	29	6	70	67	73	89	2.9	3.3	—	—	S	—	—	5,359	57	1	100	100
Barbados	0.3	14	8	0.6	-1	0.3	0.3	-1	14.2	1.7	22	12	72	70	74	50	1.6	1.5	—	72	L	—	—	166	1,626	z	99	100
Cuba	11.3	11	7	0.4	-3	11.8	11.1	-2	5.8	1.5	20	11	77	75	79	76	0.1	0.1	73	72	S	—	—	42,803	263	15	99	95
Dominica	0.1	15	7	0.8	-16	0.1	0.1	19	22.2	1.9	28	8	74	71	77	71	—	—	—	—	S	5,560	—	290	238	4	36	75
Dominican Republic	9.0	23	6	1.7	-3	11.6	14.2	57	31	2.8	33	5	68	66	69	64	1.2	1.1	70	66	H	7,150	11	18,815	479	33	67	43
Grenada	0.1	19	7	1.2	-15	0.1	0.1	-3	17	2.1	32	5	71	—	—	39	—	—	54	49	H	7,260	—	131	754	z	96	97
Guadeloupe	0.5	16	6	1.0	-1	0.5	0.5	5	7.9	2.2	26	9	78	75	82	100	—	—	—	—	—	—	—	660	698	3	64	61
Haiti	8.5	36	13	2.3	-3	13.0	18.9	121	73	4.7	42	3	52	51	54	36	3.8	3.8	28	22	H	1,840	78	10,714	795	z	52	23
Jamaica	2.7	19	6	1.3	-7	3.0	3.4	27	24	2.3	31	7	71	69	73	52	1.5	1.5	66	63	H	4,110	13	4,243	628	14	90	68
Martinique	0.4	14	7	0.7	-1	0.4	0.4	-11	6	2.0	22	12	79	76	82	95	—	—	—	—	—	—	—	425	937	11	—	—
Netherlands Antilles	0.2	13	8	0.5	21	0.2	0.2	14	9	2.0	23	10	76	72	79	69	—	—	—	—	S	—	—	309	625	1	—	—
Puerto Rico	3.9	13	7	0.6	-1	4.1	3.8	-4	8.6	1.8	22	12	77	73	81	94	—	—	78	68	—	—	—	3,456	1,137	2	—	—
St. Kitts-Nevis	0.05	18	9	1.0	-6	0.1	0.1	34	15	2.4	29	9	70	68	72	33	—	—	—	—	S	12,500	—	139	338	10	96	96
Saint Lucia	0.2	20	5	1.5	2	0.2	0.2	41	15.6	2.2	28	7	74	72	77	28	—	—	47	—	H	5,980	—	239	698	2	89	89
St. Vincent & the Grenadines	0.1	18	7	1.1	-8	0.1	0.1	-13	18.1	2.1	31	6	71	68	74	45	—	—	54	—	S	6,460	—	151	737	1	96	96
Trinidad and Tobago	1.3	14	8	0.6	-3	1.3	1.2	-6	18.6	1.6	25	7	70	67	73	74	2.6	2.6	—	—	S	13,170	—	1,981	660	2	100	100
SOUTH AMERICA	378	21	6	1.4	-1	465	528	40	25	2.4	29	6	72	69	76	80	0.5	0.5	75	66		8,210	23	6,898,579	55	19	83	42
Argentina	39.0	18	8	1.1	-1	46.4	53.7	38	16.8	2.4	27	10	74	71	78	89	0.6	0.6	—	—	S	13,920	23	1,073,514	36	6	83	—
Bolivia	9.1	31	8	2.2	-2	12.1	14.5	59	54	3.8	39	4	64	62	66	63	0.1	0.1	58	35	S	2,740	42	424,162	21	20	58	23
Brazil	186.8	21	6	1.4	0	228.9	259.8	39	27	2.3	28	6	72	68	76	81	0.5	0.5	76	70	S	8,230	21	3,300,154	57	19	83	35
Chile	16.4	16	5	1.0	0	19.1	20.2	23	7.8	2.0	25	8	78	75	81	87	0.3	0.3	—	—	S	11,470	10	292,135	56	21	96	64
Colombia	46.8	20	6	1.5	-1	58.3	66.3	42	19	2.4	31	5	72	69	75	75	0.5	0.6	78	68	H	7,420	18	439,734	106	32	96	54
Ecuador	13.3	27	6	2.1	-4	17.5	20.4	54	29	3.2	33	6	74	71	77	61	0.3	0.3	73	59	S	4,070	37	109,483	121	19	80	59
French Guiana	0.2	31	4	2.7	10	0.3	0.4	87	10	3.9	35	4	75	72	79	75	—	—	73	—	S	—	—	34,749	6	5	85	57
Guyana	0.7	22	9	1.3	-11	0.7	0.5	-35	46	2.3	36	4	76	72	80	36	2.4	2.4	37	36	S	4,230	—	83,000	9	2	86	60
Paraguay	6.3	22	5	1.7	-1	8.6	10.3	63	29	2.9	32	4	71	69	73	57	0.4	0.4	73	61	H	4,970	33	157,046	40	6	94	58
Peru	28.4	19	6	1.3	-2	34.1	35.9	27	33	2.4	31	6	70	67	72	73	0.5	0.6	71	47	H	5,830	32	496,224	57	13	72	33
Suriname	0.5	21	6	1.4	-7	0.5	0.5	-5	20	2.5	31	6	69	66	73	74	1.7	1.9	42	41	S	—	—	63,039	8	12	99	76
Uruguay	3.3	15	10	0.5	-3	3.5	3.7	12	15.3	2.2	24	13	75	71	79	93	0.4	0.5	—	—	L	9,810	6	68,498	48	z	95	85
Venezuela	27.0	22	5	1.7	0	35.2	41.7	54	17.5	2.7	31	5	73	70	76	88	0.6	0.7	—	—	S	6,440	28	352,143	77	63	71	48

© 2006 Population Reference Bureau

See Notes on page 5.

2006 World Population Data Sheet 7

Demographic Data and Estimates for the Countries and Regions of the World

	Population mid-2006 (millions)	Births per 1,000 Population	Deaths per 1,000 Pop.	Rate of Natural Increase (percent)	Net Migration per 1,000 Pop.	Projected Population (millions) mid-2025	Projected Population (millions) mid-2050	Projected Pop. Change 2006-2050 (%)	Infant Mortality Rate	Total Fertility Rate	Percent of Pop. <15	Percent of Pop. 65+	Life Expectancy Total	Life Exp. Males	Life Exp. Females	% Urban	HIV/AIDS 2003	HIV/AIDS 2005	Contraception All Methods	Contraception Modern Methods	Govt. View of Birth Rate	GNI PPP per Capita 2005	% Pop. Below US$2/Day	Area (Square Miles)	Pop. Density per Sq Mile	% Surface Area Protected 2006	Sanitation Urban	Sanitation Rural
ASIA	3,968	20	7	1.2	-0	4,739	5,277	33	49	2.4	29	6	68	66	70	38	0.4	0.4	65	59		5,960	59	12,262,691	324	11	74	31
ASIA (Excl. China)	2,657	23	7	1.6	-0	3,263	3,840	45	54	2.8	33	6	66	64	68	39	0.6	0.6	54	45		5,640	66	8,566,591	310	10	77	33
WESTERN ASIA	218	26	6	2.0	2	296	383	76	42	3.4	34	5	69	67	71	62			51	32		7,500		1,823,873	119	18	94	50
Armenia	3.0	13	9	0.4	-3	3.4	3.4	12	26	1.7	22	11	71	67	75	64	0.1	0.1	53	20	L	5,060	31	11,506	262	10	96	61
Azerbaijan	8.5	17	6	1.1	0	9.7	11.6	37	9	2.0	24	7	72	70	75	52	<0.1	0.1	55	12	S	4,890	<2	33,436	254	7	73	36
Bahrain	0.7	21	3	1.8	0	1.0	1.2	56	10	2.6	28	3	74	73	75	100			65	—	H	21,290	—	266	2,793	1	100	100
Cyprus	1.0	11	7	0.4	16	1.1	1.1	4	5	1.5	20	11	78	75	80	66			—	—	L	22,230	—	3,571	290	4	100	100
Georgia	4.4	12	11	0.1	-9	3.9	3.0	-33	25	1.6	19	13	72	69	75	52	0.1	0.2	47	27	L	3,270	25	26,911	165	4	96	69
Iraq	29.6	36	10	2.6	0	44.7	63.7	116	88	4.8	42	3	59	57	60	68			44	25	S	—	—	169,236	175	z	95	48
Israel	7.2	21	5	1.5	2	9.3	11.0	52	4.2	2.8	28	10	80	78	82	91			—	—	H	25,280	—	8,131	890	16	100	—
Jordan	5.6	29	5	2.4	2	7.9	9.9	75	24	3.7	37	4	72	71	72	82			56	41	H	5,280	7	34,444	164	11	94	85
Kuwait	2.7	19	2	1.7	14	3.9	5.1	91	10	2.4	26	2	78	77	79	96			52	39	L	24,010	—	6,880	387	3	100	87
Lebanon	3.9	19	5	1.5	-2	4.6	5.0	30	17	2.4	27	8	72	70	74	87	0.1	0.1	63	40	H	5,740	—	4,015	963	1	100	87
Oman	2.6	24	4	2.0	12	3.1	3.9	50	10	3.4	33	3	74	73	75	71			24	18	H	14,680	—	82,031	31	11	97	61
Palestinian Territory	3.9	37	4	3.3	—	7.1	11.2	188	21	5.6	46	3	72	71	74	57			51	37	H	—	—	2,417	1,609	—	78	70
Qatar	0.8	18	2	1.6	40	1.2	1.5	86	9	2.8	23	1	73	71	76	100			43	32	S	—	—	4,247	196	1	100	100
Saudi Arabia	24.1	30	3	2.7	2	35.6	47.4	96	23	4.5	38	3	72	70	74	86			32	29	S	14,740	—	829,996	29	37	100	100
Syria	19.5	29	4	2.5	0	28.1	35.9	84	18	3.5	37	3	73	71	75	50			47	35	S	3,740	—	71,498	273	2	97	56
Turkey	73.7	19	6	1.3	0	86.0	90.5	23	25	2.2	29	6	71	69	74	59			71	43	S	8,420	19	299,158	246	4	94	62
United Arab Emirates	4.9	15	1	1.3	54	7.1	9.6	94	9	2.2	25	1	77	75	80	74			28	24	S	24,090	—	32,278	153	4	100	100
Yemen	21.6	41	9	3.2	-1	38.8	68.1	214	75	6.2	46	4	60	59	62	26			23	13	H	920	45	203,849	106	0	76	14
SOUTH CENTRAL ASIA	1,642	25	8	1.7	-1	2,053	2,490	52	64	3.1	36	4	63	62	64	30	0.6	0.7	51	43		3,330	75	4,157,320	395	5	66	25
Afghanistan	31.1	48	22	2.6	0	50.3	81.9	164	166	6.8	45	2	42	41	42	22	<0.1	<0.1	10	9	H	—	—	251,772	123	z	16	5
Bangladesh	146.6	27	8	1.9	-0	190.0	231.0	58	65	3.0	35	3	61	61	62	23	<0.1	<0.1	58	47	H	2,090	83	55,598	2,637	1	75	39
Bhutan	0.9	20	7	1.3	0	1.3	1.8	99	40	2.9	33	5	63	62	64	31	<0.1	<0.1	—	31	H	—	—	18,147	49	26	65	70
India	1,121.8	24	8	1.7	-0	1,363.0	1,628.0	45	58	2.9	36	4	62	62	63	29	0.9	0.9	53	46	H	3,460	80	1,269,340	884	5	58	18
Iran	70.3	18	6	1.2	-4	89.0	101.9	45	32	2.0	29	5	70	69	72	67	0.1	0.2	74	56	H	8,050	7	630,575	112	7	86	78
Kazakhstan	15.3	18	10	0.8	1	16.0	15.2	-1	29	2.2	27	8	66	61	72	57	0.1	0.1	66	53	L	7,730	16	1,049,151	15	3	87	52
Kyrgyzstan	5.2	21	7	1.4	-5	6.6	8.2	58	30	2.6	32	6	68	64	72	35	<0.1	0.1	60	49	S	1,870	21	76,641	67	4	75	51
Maldives	0.3	18	3	1.5	0	0.4	0.5	80	15	2.8	33	5	70	70	70	27			40	35	H	—	—	116	2,573	0	100	42
Nepal	26.0	31	9	2.2	-1	36.2	48.0	85	64	3.7	41	4	62	62	63	14	0.5	0.5	39	35	H	1,530	69	56,826	457	16	68	20
Pakistan	165.8	33	9	2.4	-1	228.8	295.0	78	79	4.6	41	4	61	61	63	34	0.1	0.1	28	20	H	2,350	74	307,375	539	9	92	35
Sri Lanka	19.9	19	6	1.3	-1	22.2	22.4	13	11	2.0	26	7	74	71	77	20	<0.1	<0.1	70	50	S	4,520	42	25,332	784	17	98	89
Tajikistan	7.0	30	8	2.2	-1	9.3	11.1	58	89	3.8	32	3	64	61	66	26	<0.1	0.1	34	27	H	1,260	43	55,251	127	18	71	47
Turkmenistan	5.3	25	8	1.6	-0	6.6	7.4	38	74	2.9	34	5	62	58	67	47	<0.1	<0.1	62	53	S	—	—	188,456	28	4	77	50
Uzbekistan	26.2	23	7	1.6	-2	33.0	37.5	43	58	2.7	35	5	67	63	70	36	0.1	0.2	68	63	S	2,020		172,741	152	5	73	48

Demographic Data and Estimates for the Countries and Regions of the World

	Population mid-2006 (millions)	Births per 1,000 Population	Deaths per 1,000 Population	Rate of Natural Increase (percent)	Net Migration per 1,000 Pop.	Projected Population (millions) mid-2025	Projected Population (millions) mid-2050	Projected Pop. Change 2006–2050 (%)	Infant Mortality Rate	Total Fertility Rate	% of Pop. of Age <15	% of Pop. of Age 65+	Life Exp. Total	Life Exp. Males	Life Exp. Females	% Urban	% HIV/AIDS 2003	% HIV/AIDS 2005	Contra. All Methods	Contra. Modern Methods	Govt. View of Birth Rate	GNI PPP per Capita 2005	% Pop. Below US$2/Day	Area (Sq. Miles)	Pop. Density per Sq. Mile	% Surface Area Protected 2006	Sanitation Urban 2002	Sanitation Rural 2002
SOUTHEAST ASIA	565	21	6	1.4	-1	682	766	36	34	2.5	30	5	69	66	71	39	0.5	0.5	60	52		4,530	46	1,735,448	326	10	79	50
Brunei	0.4	20	3	1.7	-1	0.5	0.7	87	9	2.4	30	3	75	72	77	72	<0.1	<0.1			S			2,228	164	38		
Cambodia	14.1	30	9	2.1	-0	19.6	25.5	81	91	3.7	37	3	60	57	63	15	2.0	1.6	24	19	S	2,490	78	69,900	201	22	53	8
East Timor	1.0	42	15	2.7	1	1.9	3.2	224	88	6.3	43	3	56	54	57	22	—	—	10	9	H			5,741	170	1	65	30
Indonesia	225.5	20	6	1.4	-1	263.7	284.6	26	35	2.4	29	5	69	67	72	42	0.1	0.1	60	57	H	3,720	52	735,355	307	9	71	38
Laos	6.1	36	13	2.3	-0	8.7	11.6	91	88	4.8	43	4	54	53	56	19	0.1	0.1	32	29	H	2,020	74	91,429	66	16	61	14
Malaysia	26.9	20	4	1.6	4	34.6	40.6	51	10	2.6	33	4	74	72	76	62	0.4	0.5	55	30	S	10,320	9	127,317	211	17	96	98
Myanmar	51.0	21	10	1.1	0	59.0	63.7	25	75	2.5	32	5	60	57	63	29	1.4	1.3	37	33	S			261,228	195	5	96	63
Philippines	86.3	27	5	2.1	-2	115.7	142.2	65	27	3.4	35	4	70	67	72	48	<0.1	<0.1	49	33	H	5,300	48	115,830	745	7	81	61
Singapore	4.5	10	4	0.6	20	5.2	5.3	19	2.1	1.2	20	8	80	78	82	100	0.3	0.3	62	55	L	29,780	—	239	18,652	2	100	—
Thailand	65.2	14	7	0.7	-0	70.2	73.2	12	20	1.7	23	7	71	68	75	33	1.4	1.4	79	79	S	8,440	25	198,116	329	19	97	100
Vietnam	84.2	19	5	1.3	-0	102.9	115.1	37	18	2.1	29	7	72	70	73	26	0.4	0.5	77	66	S	3,010	—	128,066	657	4	84	26
EAST ASIA	1,544	12	7	0.5	-0	1,708	1,639	6	25	1.6	20	9	73	71	75	43	0.1	0.1	84	82	S	9,050	47	4,545,050	340	14	74	32
China	1,311.4	12	7	0.6	-0	1,476.0	1,437.0	10	27	1.6	20	8	72	70	74	37	0.1	0.1	87	86	S	6,600	47	3,696,100	355	15	69	29
China, Hong Kong SAR[d]	7.0	8	6	0.3	5	8.1	8.7	24	2.4	1.0	14	12	81	79	84	100	—	—	—	—		34,670	—	413	16,915	51	—	—
China, Macao SAR[d]	0.5	8	3	0.4	44	0.6	0.6	12	3	0.9	16	8	79	77	82	99	—	—	—	—				8	64,745	—	—	—
Japan	127.8	9	8	0.0	0	121.1	100.6	-21	2.8	1.3	14	20	82	79	86	79	<0.1	<0.1	56	48	L	31,410	—	145,869	876	9	100	100
Korea, North	23.1	16	7	0.9	0	25.8	26.4	14	21	2.0	27	8	71	68	73	60	—	—	69	58	L			46,541	497	2	58	60
Korea, South	48.5	9	5	0.4	-1	49.8	42.3	-13	5	1.1	19	10	77	74	81	82	<0.1	<0.1	81	67	L	21,850	<2	38,324	1,265	4	58	37
Mongolia	2.6	18	6	1.2	0	3.1	3.5	35	21	1.9	29	4	66	64	68	57	<0.1	<0.1	67	54	L	2,190	75	604,826	4	14	75	37
Taiwan	22.8	9	6	0.3	1	23.6	19.8	-13	5.4	1.1	19	10	76	73	79	78	—	—	71	—				13,969	1,633	6	—	—
EUROPE	732	10	12	-0.1	2	717	665	-9	7	1.4	16	16	75	71	79	75	0.5	0.5	68	53	S	21,120	—	8,875,867	82	9	—	—
NORTHERN EUROPE	97	12	10	0.2	4	103	107	11	5	1.7	18	16	78	75	81	82	0.2	0.2	82	76	S	31,570	—	675,794	143	9	—	—
Channel Islands	0.2	12	9	0.2	—	0.2	0.2	15	3.4	1.4	15	13	78	76	80	31	—	—	—	—				75	1,987	—	—	—
Denmark	5.4	12	10	0.2	1	5.6	5.5	1	4.4	1.8	19	15	78	76	80	72	0.2	0.2	—	56	S	33,570	—	16,637	327	7	—	—
Estonia	1.3	11	13	-0.2	-0	1.2	1.0	-23	6	1.5	15	17	72	66	78	69	1.1	1.3	70	56	L	15,420	8	17,413	77	31	93	—
Finland	5.3	11	9	0.2	1	5.4	5.3	0	3.0	1.8	17	16	79	75	82	62	0.1	0.1	79	78	L	31,170	—	130,560	40	8	100	100
Iceland	0.3	14	6	0.8	5	0.3	0.4	17	2.5	2.1	22	12	81	79	83	93	0.2	0.2	—	—	L	34,760	—	39,768	8	6	100	100
Ireland	4.2	15	7	0.8	13	4.5	4.7	11	4.7	1.9	21	11	78	75	80	60	0.2	0.2	—	—	S	34,720	—	27,135	156	1	—	—
Latvia	2.3	9	14	-0.5	-0	2.2	1.8	-23	7	1.3	15	17	73	67	77	68	0.6	0.8	85	60	L	13,480	5	24,942	92	14	—	—
Lithuania	3.4	9	13	-0.4	-3	3.1	2.9	-15	7	1.3	17	15	72	66	78	67	0.1	0.2	47	30	L	14,220	8	25,174	135	11	—	—
Norway	4.7	12	9	0.3	4	5.2	5.8	25	3.1	1.8	20	15	80	78	83	78	0.1	0.1	—	—	S	40,420	—	125,050	37	5	100	100
Sweden	9.1	11	10	0.1	4	9.9	10.5	16	2.4	1.8	17	17	81	78	83	84	0.2	0.2	—	—	S	31,420	—	173,730	52	9	100	100
United Kingdom	60.5	12	10	0.2	4	65.8	69.2	14	5.1	1.8	18	16	79	76	81	89	0.2	0.2	84	79	S	32,690	—	94,548	640	13	100	100
WESTERN EUROPE	187	10	9	0.1	2	190	184	-2	4	1.6	16	17	79	76	82	80	0.2	0.2	74	70	S	30,690	—	427,702	437	19	—	—
Austria	8.3	9	9	0.0	7	8.7	9.0	8	4.1	1.4	16	16	79	76	82	54	0.3	0.3	67	65	L	33,140	—	32,378	256	28	100	100
Belgium	10.5	11	10	0.1	3	10.8	11.0	4	4.8	1.6	17	17	79	76	82	97	0.3	0.3	79	75	S	32,640	—	11,787	893	3	—	—
France	61.2	13	9	0.4	2	63.4	64.0	5	3.6	1.9	18	16	80	77	84	76	0.4	0.4	75	69	L	30,540	—	212,934	287	12	—	—
Germany	82.4	8	10	-0.2	1	82.0	75.1	-9	3.9	1.3	14	19	79	76	82	88	0.1	0.1	75	72	L	29,210	—	137,830	598	30	100	100
Liechtenstein	0.04	11	6	0.5	2	0.04	0.04	26	2.9	1.4	18	11	80	79	82	21	—	—	—	—	L			62	567	40	—	—
Luxembourg	0.5	12	8	0.4	8	0.5	0.6	37	3.9	1.7	19	14	79	75	81	91	0.2	0.2	—	—	S	65,340	—	999	460	17	—	—
Monaco	0.03	23	16	0.6	8	0.04	0.1	67	—	—	13	22	—	—	—	100	—	—	—	—	S			1	44,000	26	100	100
Netherlands	16.4	12	8	0.3	-2	16.9	16.9	3	4.9	1.7	18	14	79	77	81	65	0.2	0.2	79	76	S	32,480	—	15,768	1,037	14	100	100
Switzerland	7.5	10	8	0.2	5	7.4	7.2	-4	4.3	1.4	16	16	81	79	84	68	0.4	0.4	57	54	L	37,080	—	15,942	469	29	100	100

See Notes on page 5.

Demographic Data and Estimates for the Countries and Regions of the World

	Population mid-2006 (millions)	Births per 1,000 Pop.	Deaths per 1,000 Pop.	Rate of Natural Increase (%)	Net Migration per 1,000 Pop.	Projected Population (millions) mid-2025	Projected Population (millions) mid-2050	Projected Pop. Change 2006-2050 (%)	Infant Mortality Rate	Total Fertility Rate	% of Pop. <15	% of Pop. 65+	Life Exp. Total	Life Exp. Males	Life Exp. Females	% Urban	% HIV/AIDS 15-49 2003	% HIV/AIDS 15-49 2005	Contraception All Methods	Contraception Modern Methods	Govt. View of Birth Rate	GNI PPP per Capita 2005	% Pop. Below US$2/Day	Area (Sq. Miles)	Pop. Density per Sq. Mile	% Surface Area Protected 2006	Sanitation Urban 2002	Sanitation Rural 2002
EASTERN EUROPE	296	10	14	-0.5	0	271	230	-22	10	1.3	16	14	69	63	74	68	0.8	0.8	64	42		10,640	9	7,254,035	41	9	94	70
Belarus	9.7	9	15	-0.6	0	9.4	8.5	-12	8	1.2	16	14	69	63	75	72	0.3	0.3	50	42	L	7,890	<2	80,154	121	6	—	—
Bulgaria	7.7	9	15	-0.5	-0	6.6	5.1	-34	10.4	1.3	14	17	72	69	76	70	—	<0.1	41	26	L	8,630	6	42,822	180	10	100	100
Czech Republic	10.3	10	11	-0.1	4	10.2	9.4	-8	3.4	1.3	15	14	76	73	79	77	<0.1	0.1	67	58	L	20,140	<2	30,448	337	16	100	100
Hungary	10.1	10	13	-0.3	0	9.6	8.9	-11	6.1	1.3	16	16	73	69	77	65	0.1	0.1	77	68	L	16,940	<2	35,919	280	9	100	85
Moldova	4.0	11	12	-0.2	1	3.8	3.1	-21	12	1.3	20	10	69	65	72	45	0.9	1.1	62	43	L	2,150	64	13,012	306	1	86	52
Poland	38.1	10	10	-0.0	-0	36.7	31.5	-17	6.4	1.3	17	13	75	71	79	62	0.1	0.1	49	19	L	13,490	<2	124,807	306	27	—	—
Romania	21.6	10	12	-0.2	-1	18.1	15.3	-29	16.8	1.3	16	14	71	68	75	55	—	<0.1	64	30	L	8,940	13	92,042	234	5	86	10
Russia	142.3	10	16	-0.6	1	130.0	110.3	-22	11	1.3	15	14	65	59	72	73	0.9	1.1	67	49	L	10,640	12	6,592,819	22	9	93	70
Slovakia	5.4	10	10	0.0	1	5.2	4.7	-12	6.8	1.3	17	12	74	70	78	56	<0.1	<0.1	74	41	L	15,760	3	18,923	285	25	100	100
Ukraine	46.8	9	17	-0.8	1	41.7	33.4	-28	10	1.2	14	16	68	63	74	68	1.3	1.4	68	38	L	6,720	5	233,089	201	3	100	97
SOUTHERN EUROPE	152	10	10	0.1	5	153	144	-6	5	1.4	15	17	79	76	82	75	0.5	0.5	59	43		23,090		508,337	300	7		
Albania	3.2	14	6	0.8	-3	3.5	3.5	12	8	1.9	27	8	75	72	79	45	—	—	75	8	S	5,420	12	11,100	284	3	99	81
Andorra	0.1	11	4	0.7	43	0.1	0.1	-2	3.9	1.3	15	12	75	72	79	92	—	—	—	—	S	—	—	174	501	7	100	100
Bosnia-Herzegovina	3.9	9	9	0.1	0	3.7	3.2	-18	7	1.3	16	14	74	71	77	43	—	<0.1	48	16	L	7,790	—	19,741	196	1	99	88
Croatia	4.4	9	11	-0.2	3	4.3	3.8	-14	6.1	1.4	16	16	75	71	78	56	<0.1	<0.1	—	—	L	12,750	<2	21,830	204	6	100	100
Greece	11.1	10	10	0.0	3	11.4	10.6	-4	4.0	1.3	15	18	79	77	81	60	0.2	0.2	60	39	L	23,620	—	50,950	218	3	—	—
Italy	59.0	10	10	-0.0	5	58.7	55.9	-5	4.1	1.3	14	19	80	78	83	90	0.5	0.5	60	39	S	28,840	—	116,320	507	13	—	—
Macedonia	2.0	11	9	0.2	-0	2.1	1.9	-7	11.3	1.4	21	11	73	71	76	59	<0.1	<0.1	59	—	S	7,080	<2	9,927	206	7	—	—
Malta	0.4	9	7	0.2	5	0.4	0.4	-10	5.9	1.4	18	13	79	77	81	91	0.1	0.1	86	43	S	18,960	—	124	3,278	—	100	100
Montenegro	0.6	13	9	0.3	0	0.6	0.6	-4	8	1.7	21	12	74	—	—	—	—	—	—	—	—	—	—	5,333	117	—	—	—
Portugal	10.6	10	10	0.1	5	10.4	9.3	-12	3.8	1.4	16	17	78	75	81	53	0.4	0.4	—	—	L	19,730	<2	35,514	299	5	—	—
San Marino	0.03	10	8	0.3	11	0.04	0.04	13	6.7	1.2	15	16	81	78	84	84	—	—	—	—	—	—	—	23	1,338	—	—	—
Serbia	9.5	13	12	0.1	1	9.2	8.5	-10	10	1.8	19	15	72	69	75	52	0.2	0.2	58	33	L	—	<2	34,115	277	4	97	77
Slovenia	2.0	9	9	-0.0	3	2.0	1.9	-5	3.9	1.3	14	15	77	74	81	49	<0.1	<0.1	71	57	L	22,160	<2	7,819	256	7	—	—
Spain	45.5	11	9	0.2	6	46.2	43.9	-4	4.0	1.3	14	17	81	77	84	76	0.7	0.6	56	53	L	25,820	—	195,363	233	8	—	—
OCEANIA	34	17	7	1.0	3	41	48	43	27	2.1	25	10	75	73	77	73	0.4	0.4	72	63		22,180		3,306,741	10	13	98	58
Australia	20.6	13	6	0.6	5	24.6	28.1	36	4.9	1.8	20	13	81	78	83	91	0.1	0.1	85	75	L	30,610	—	2,988,888	7	17	100	100
Fed. States of Micronesia	0.1	26	6	2.0	-19	0.1	0.1	-10	40	4.1	39	3	67	67	67	22	—	—	—	—	H	—	—	270	400	z	61	14
Fiji	0.8	21	5	1.4	-5	0.9	0.9	9	16	2.5	31	4	68	66	71	46	0.1	0.1	—	—	S	5,960	—	7,054	120	z	99	98
French Polynesia	0.3	18	5	1.3	2	0.3	0.4	39	5.2	2.2	29	5	74	72	77	53	—	—	—	—	—	—	—	1,544	168	z	99	97
Guam	0.2	21	4	1.6	0	0.2	0.2	42	11.2	2.7	30	6	78	75	81	93	—	—	—	—	—	—	—	212	805	z	99	98
Kiribati	0.1	31	8	2.3	-1	0.1	0.2	123	43	4.2	39	3	61	58	64	43	—	—	—	—	H	—	—	282	334	59	59	22
Marshall Islands	0.1	38	5	3.3	-6	0.1	0.1	65	29	4.9	42	2	70	—	—	68	—	—	—	—	H	—	—	69	935	93	93	59
Nauru	0.01	26	7	1.9	0	0.02	0.02	77	42	3.4	39	2	62	58	66	100	—	—	—	—	S	—	—	9	1,529	—	—	—
New Caledonia	0.2	17	5	1.2	7	0.3	0.4	59	6	2.2	28	6	74	71	77	71	—	—	—	—	S	—	—	7,174	33	2	—	—
New Zealand	4.1	14	7	0.7	2	4.6	4.9	18	5.1	2.0	21	12	79	77	81	89	0.1	0.1	74	72	S	23,030	—	104,452	40	20	—	—
Palau	0.02	14	7	0.7	1	0.02	0.03	30	18	2.1	24	5	71	69	73	77	—	—	—	—	S	—	—	178	113	z	96	52
Papua New Guinea	6.0	32	11	2.1	0	8.2	10.6	77	64	4.1	41	2	55	55	56	13	1.6	1.8	26	20	H	2,370	—	178,703	34	4	67	41
Samoa	0.2	29	6	2.4	-1	0.2	0.2	-15	20	4.4	41	4	73	72	74	22	—	—	—	—	H	6,480	—	1,097	170	z	100	100
Solomon Islands	0.5	34	8	2.6	0	0.7	1.1	120	48	4.5	40	3	62	62	63	16	—	—	—	—	H	1,880	—	11,158	43	18	98	18
Tonga	0.1	25	7	1.8	-14	0.1	0.2	66	19	3.1	35	6	71	70	72	23	—	—	—	—	S	8,040	—	290	356	28	98	96
Tuvalu	0.01	27	10	1.7	-1	0.01	0.02	80	35	3.7	36	6	64	62	65	47	—	—	—	—	H	—	—	10	1,000	z	92	83
Vanuatu	0.2	31	6	2.5	—	0.4	0.4	89	27	4.0	41	3	67	66	69	21	—	—	—	20	H	3,170	—	4,707	48	z	78	42

Acknowledgments, Notes, Sources, and Definitions

Acknowledgments

The author gratefully acknowledges the valuable assistance of PRB staff members Lori Ashford, Donna Clifton, Zuali Malsawma, and Kelvin Pollard; staff of the International Programs Center of the U.S. Census Bureau; the United Nations (UN) Population Division; the Institut national d'etudes démographiques (INED), Paris; and the World Bank in the preparation of this year's *World Population Data Sheet*. This publication is funded in part by the U.S. Agency for International Development (Cooperative Agreement GPO-A-oo-o3-oooo4-oo) and by PRB members and supporters.

The information in this data sheet was not provided by and does not represent the views of the United States government or the USAID.

Notes

The *Data Sheet* lists all geopolitical entities with populations of 150,000 or more and all members of the UN. These include sovereign states, dependencies, overseas departments, and some territories whose status or boundaries may be undetermined or in dispute. **More developed regions**, following the UN classification, comprise all of Europe and North America, plus Australia, Japan, and New Zealand. All other regions and countries are classified as **less developed**.

Sub-Saharan Africa: All countries of Africa except the northern African countries of Algeria, Egypt, Libya, Morocco, Tunisia, and Western Sahara.

World and Regional Totals: Regional population totals are independently rounded and include small countries or areas not shown. Regional and world rates and percentages are weighted averages of countries for which data are available; regional averages are shown when data or estimates are available for at least three-quarters of the region's population.

World Population Data Sheets from different years should **not be used as a time series.** Fluctuations in values from year to year often reflect revisions based on new data or estimates rather than actual changes in levels. Additional information on likely trends and consistent time series can be obtained from PRB, and are also available in UN and U.S. Census Bureau publications.

Sources

The rates and figures are primarily compiled from the following sources: official country statistical yearbooks and bulletins; *United Nations Demographic Yearbook, 2001* of the UN Statistics Division; *World Population Prospects: The 2004 Revision* of the UN Population Division; the UN Statistical Library; *Recent Demographic Developments in Europe, 2004* of the Council of Europe; Country, Regional and Global Estimates on Water and Sanitation of UNICEF and World Health Organization;

World Database on Protected Areas of UNEP World Conservation Monitoring Center; and the data files and library resources of the International Programs Center, U.S. Census Bureau. Other sources include recent demographic surveys such as the Demographic and Health Surveys, Reproductive Health Surveys, special studies, and direct communication with demographers and statistical bureaus in the United States and abroad. Specific data sources may be obtained by contacting the author of the *2006 World Population Data Sheet*.

For countries with complete registration of births and deaths, rates are those most recently reported. For more developed countries, nearly all vital rates refer to 2004 or 2005, and for less developed countries, for some point in the early to mid-2000s.

Definitions

Mid-2006 Population

Estimates are based on a recent census, official national data, or UN and U.S. Census Bureau projections. The effects of refugee movements, large numbers of foreign workers, and population shifts due to contemporary political events are taken into account to the extent possible.

Birth and Death Rate

The annual number of births and deaths per 1,000 total population. These rates are often referred to as "crude rates" since they do not take a population's age structure into account. Thus, crude death rates in more developed countries, with a relatively large proportion of high-mortality older population, are often higher than those in less developed countries with lower life expectancy.

Rate of Natural Increase (RNI)

The birth rate minus the death rate, implying the annual rate of population growth without regard for migration. Expressed as a percentage.

Net Migration

The estimated rate of net immigration (immigration minus emigration) per 1,000 population for a recent year based upon the official national rate or derived as a residual from estimated birth, death, and population growth rates. Migration rates can vary substantially from year to year for any particular country.

Projected Population 2025 and 2050

Projected populations based upon reasonable assumptions on the future course of fertility, mortality, and migration. Projections are based upon official country projections, series issued by the UN or the U.S. Census Bureau, or PRB projections.

Infant Mortality Rate

The annual number of deaths of infants under age 1 per 1,000 live births. Rates shown with decimals indicate national statistics reported as completely registered, while those without are estimates from the sources cited above. Rates shown in italics are based upon fewer than 50 annual infant deaths and, as a result, are subject to considerable yearly variability.

Total Fertility Rate (TFR)

The average number of children a woman would have assuming that current age-specific birth rates remain constant throughout her childbearing years (usually considered to be ages 15 to 49).

Population Under Age 15/Age 65+

The percentage of the total population in these ages, which are often considered the "dependent ages."

Life Expectancy at Birth

The average number of years a newborn infant can expect to live under current mortality levels.

Percent Urban

Percentage of the total population living in areas termed "urban" by that country. Typically, the population living in towns of 2,000 or more or in national and provincial capitals is classified "urban."

Percent of Adult Population Ages 15 to 49 With HIV/AIDS

The estimated percentage of adults living with HIV/AIDS in 2003 and 2005. Nearly all data are from UNAIDS' *2006 Report on the Global AIDS Epidemic.*

Contraceptive Use

The percentage of currently married or "in-union" women of reproductive age who are currently using any form of contraception.

"Modern" methods include clinic and supply methods such as the pill, IUD, condom, and sterilization. Data are from the most recent available national-level surveys, such as the Demographic and Health Surveys, Reproductive Health Survey programs, and the UN Population Division *World Contraceptive Use 2003*. Other sources include direct communication with national statistical organizations and the databases of the UN Population Division and the U.S. Census Bureau. Data refer to some point in the 1990s and early 2000s. Data prior to 2000 are shown in italics.

Government View of Current Birth Rate

This population policy indicator presents the officially stated position of country governments on the level of the national birth rate. Indicators are from the UN Population Division, *World Population Policies 2005.*

GNI PPP per Capita, 2005 (US$)

GNI PPP per capita is gross national income in purchasing power parity (PPP) divided by midyear population. GNI PPP refers to gross national income converted to "international" dollars using a purchasing power parity conversion factor. International dollars indicate the amount of goods and services one could buy in the United States with a given amount of money. Data are from the World Bank. Figures in italics are for 2003 or 2004.

Percent of Population Living Below US$2/Day

The proportion of people living below $2 per day is the percentage of the population with average consumption expenditures less than $2.15 per day measured in 1993 prices converted using purchasing power parity (PPP) rates. The World Bank's estimates are drawn from surveys that use common methods for measuring household living standards across countries. When estimating poverty worldwide, the same reference poverty line has to be used, expressed in a common unit across countries. The World Bank uses reference lines set at $1 per day (extreme poverty) and $2 per day (poverty) in 1993 PPP terms, where PPPs measure the relative purchasing power of currencies across countries. For analysis of poverty trends in a particular country, use of the national poverty line is preferable. Most data refer to the late 1990s and early 2000s.

Population With Access to Improved Sanitation (%)

The percentage of the population using improved sanitation facilities. Improved facilities are those more likely to ensure privacy and hygienic use. Improved facilities include connection to a public sewer, connection to a septic system, pour-flush latrines, simple pit latrines, and/or ventilated improved pit latrines. Unimproved facilities include public or shared atrines, open pit latrines, or bucket latrines.

Surface Area Protected (%)

The percentage of a country's total surface area nationally designated as protected under one of several categories designated by the World Conservation Union (IUCN). The categories are: a strict nature reserve, a wilderness area, a national park, a natural monument, a habitat or species management area, a protected landscape or seascape, and/or a managed resource protected area, as well as nationally designated protected areas for which no IUCN category has been defined or provided by a national agency. The total surface area of a country includes terrestrial area plus any territorial sea area (up to 12 nautical miles).

The **POPULATION REFERENCE BUREAU informs** people around the world about population, health, and the environment, and **empowers** them to use that information to **advance** the well-being of current and future generations.

inform

PRB informs people around the world and in the United States about issues related to population, health, and the environment. To do this, we transform technical data and research into accurate, easy-to-understand information.

Innovative Tools. PRB's wallcharts, including the *World Population Data Sheet* and the *Map of Persistent Child Poverty in the U.S.*, are searchable via our DataFinder web tool and make accurate demographic information accessible to a wide audience.

Influential Reports. Health workers in the developing world use PRB's report on cervical cancer prevention, created in collaboration with the global health nonprofit PATH, to design successful screening programs. PRB and the Russell Sage Foundation published *The American People: Census 2000*, 14 reports that describe America in the year 2000.

Unbiased Policy Analysis. For more than 20 years, PRB has hosted a monthly seminar series focused on the policy implications of population issues including the color line in American society and HIV/AIDS in India.

Online Resources. PRB's website offers full text of all PRB publications, including our respected *Population Bulletins* and web-exclusive data and analysis on world issues ranging from aging to family planning. Our Center for Public Information on Population Research puts new population research findings into context for journalists and policymakers.

empower

PRB empowers people—researchers, journalists, policymakers, and educators—to use information about population, health, and the environment to encourage action.

Information alone can be powerful. Frequently, however, people have knowledge but lack the tools needed to communicate it effectively to decisionmakers. PRB builds coalitions and conducts trainings in the United States and throughout the developing world to share techniques to inform policy.

Journalist Networks. Since 1996, PRB has shared techniques for fact-based, reproductive health reporting with a network of West African editors. The Pop'Médiafrique program, one of several PRB journalist networks, has improved news coverage and increased demand for family planning in the region.

Policy Communications Training. Over the past five years, PRB has trained nearly 500 advocates, health professionals, and government workers in Asia, Africa, and Latin America. For example, participants in a workshop in Madagascar learned how to develop a fact sheet for policymakers to explain the complex linkages between population, health, and the environment.

Data Workshops. PRB's data workshops assist the Annie E. Casey Foundation's KIDS COUNT network in using vital data about the status of children in the United States. Workshop participants take away the knowledge needed to access data about their particular state and communicate with policymakers.

advance

PRB works to advance the well-being of current and future generations. Toward that end, we analyze data and research, disseminate information, and empower people to use that information in order to inform policymaking.

While the numbers of publications created or workshops conducted are one way to measure PRB's work, the creation of evidence-based policies, increased demand for health services, and active coalitions are better gauges of progress toward positive social change.

Evidence-Based Policies. PRB provides analysis for the *KIDS COUNT Data Book*, an annual report card on the well-being of children and families in the United States, that has helped promote the passage of several U.S. policies, including the State Children's Health Insurance Program.

Increased Demand for Health Services. Information broadcast by women radio journalists who attended PRB's reproductive health workshop in Senegal has increased local demand for family planning and health services.

Active Coalitions. PRB worked with local groups in the Philippines to establish a national coalition that helps decisionmakers understand the impact of population on the environment through events such as an International Earth Day celebration near the endangered Pasig River in Manila.

For a full list of PRB publications available in English, French, Spanish, Arabic, and Portuguese, go to PRB's online store at www.prb.org.
  To order PRB publications (discounts available for bulk orders):
• Online at www.prb.org.
• E-mail: popref@prb.org.
• Call toll-free: 800-877-9881.
• Fax: 202-328-3937.
• Mail: 1875 Connecticut Ave., NW, Suite 520, Washington, DC 20009.

The 2006 *World Population Data Sheet* is also available in French and Spanish.

© August 2006 Population Reference Bureau.
ISSN 0085-8315.
Data prepared by PRB demographer Carl Haub.
Graphs and tables prepared by PRB demographer Kelvin Pollard.
Design and production: Michelle Corbett, PRB.

PRB's *World Population Data Sheet* is used around the world and is widely considered to be the most accurate source of information on population. If you value the *Data Sheet* and are among the tens of thousands of people who eagerly anticipate its publication each year, please consider making a contribution to PRB. Your donation will help ensure that PRB can maintain its commitment to keeping the *Data Sheet* as affordable as possible. Visit our website to donate now: www.prb.org.

POPULATION REFERENCE BUREAU

1875 Connecticut Ave. NW, Washington, DC 20009 USA
tel. 202-483-1100 | fax 202-328-3937
email: popref@prb.org | website: www.prb.org